62nd International Scientific Conference on Experimental Stress Analysis (EAN 2024)

Book of Abstracts

Bozi Dar, Czech Republic
4-6 June 2024

ISBN: 979-8-3313-1105-6

Printed from e-media with permission by:

Curran Associates, Inc.
57 Morehouse Lane
Red Hook, NY 12571

CURRAN ASSOCIATES INC. proceedings .com

Some format issues inherent in the e-media version may also appear in this print version.

Czech Society for Mechanics (CSM)
Dolejskova 5
Praha 8
182 00
Czech Republic

Phone: +420 266 053 045
Fax: +420 286 587 784

csm@csm.cz

Additional copies of this publication are available from:

Curran Associates, Inc.
57 Morehouse Lane
Red Hook, NY 12571 USA
Phone: 845-758-0400
Fax: 845-758-2633
Email: curran@proceedings.com
Web: www.proceedings.com

Experimental Stress Analysis 2024

62nd International Conference
June 4th – 6th, 2024, Boží Dar, Czech Republic

EAN 2024
62nd conference on experimental stress analysis

Book of Extended Abstracts

EDITED BY

PAVEL POLACH
LUKÁŠ STUNA

Výzkumný a zkušební ústav Plzeň s.r.o.
(Research and Testing Institute Plzen)

Czech Society for Mechanics
expert group for Experimental Mechanics

The editors did their best to ensure the printing quality of the proceedings. However, some manuscripts did not comply with the proceeding's instructions. Therefore, the editors cannot accept responsibility for the final appearance of these manuscripts.

Authors: Collective Authorship

Published by:
Západočeská univerzita v Plzni / University of West Bohemia in Pilsen
Univerzitní 8, 301 00 Plzeň, Czech Republic

Printed by:
BÍLÝ SLON - COPYDIGITAL s.r.o.
Škroupova 1017/11, 301 00 Plzeň

First edition, 2024

This page intentionally left blank.

Main Subject Areas

- Development of experimental methods in mechanics
- Development of experimental methods in biomechanics
- New methods and applications of deformation and stress analysis in mechanical structures
- New methods and applications of deformation and stress analysis in civil and other structures
- Experiment as a tool for the verification of analytical and numerical methods
- Experimental research and prediction of strength, life, and operational reliability of structures
- Monitoring of operational loadings and operational states of structures and devices
- Mechanics of advanced materials, composites and 3D printing materials
- Methods and means of teaching experimental methods

Program Committee

Members

- Karel Doubrava – Czech Technical University in Prague, CZE
- Peter Frankovský – Technical University of Košice, SVK
- Nikolaj Ganev – Czech Technical University in Prague, CZE
- Radim Halama – Technical University of Ostrava, CZE
- František Holešovský – University of West Bohemia, CZE
- Stanislav Holý – Czech Technical University in Prague, CZE
- Lubomír Houfek – Brno University of Technology, CZE
- Miroslav Hrabovský – Palacký University in Olomouc, CZE
- Róbert Huňady – Technical University of Košice, SVK
- Josef Jíra – Czech Technical University in Prague, CZE
- Kamil Kolařík – Czech Technical University in Prague, CZE
- Aleš Lufinka – Technical University of Liberec, CZE
- Tomáš Návrat – Brno University of Technology, CZE
- Pavel Padevět – Czech Technical University in Prague, CZE
- Iva Petríková – Technical University of Liberec, CZE
- Roman Růžek – Czech Aerospace Research Centre, CZE
- Milan Růžička – Czech Technical University in Prague, CZE
- Jaroslav Václavík – Research and Testing Institute Plzen, CZE
- Robert Zemčík – University of West Bohemia, CZE

Organizing Committee

Members

- Lenka Lopatková
- Radka Musilová
- Pavel Polach
- Jaroslav Václavík

Conference Partners

DEWESoft®

Gantner instruments

HBM HBP měřicí technika

KESACON
- official Kyowa representative -

LABTECH

MICRO-EPSILON

SHIMADZU
Excellence in Science

YOTTAVOLT

Table of Contents

X-ray Computed Tomography of the Oscillating Beam

P. Beneš [1,a], T. Fíla [1], P. Koudelka [1], D. Kytýř [1], M. Macháček [1], V. Rada [1], P. Zlámal [1], D. Vavřík [1]

[1] *Institute of Theoretical and Applied Mechanics, Czech Academy of Sciences, Prosecká 76, 190 00, Prague 9, Czech Republic*

[a] *benes@itam.cas.cz*

Abstract: In the field of experimental mechanics, the trend of using four-dimensional computed tomography (4D CT) to monitor various time-dependent processes has been gaining momentum in recent years. The 4D CT method makes it possible to obtain data concerning the structural and geometrical changes of the object under study caused by external loading. In this particular case study, a modal analysis of the beam oscillation at its first eigenfrequency will be presented.

Keywords: computed tomography; modal analysis; physical process monitoring.

Extended Abstract

In the field of experimental mechanics, the trend of using four-dimensional computed tomography (4D CT) to monitor various time-dependent processes has been gaining momentum in recent years. The 4D CT method makes it possible to obtain data concerning the structural and geometrical changes of the object under study caused by external loading. In this particular case study, a modal analysis of the beam oscillation at its first eigenfrequency will be presented.

The number of X-ray projections required for a complete tomographic reconstruction during the relative rotation of an object can range from hundreds to thousands, depending on the type of object and the loading. With traditional CT scanners, a temporal resolution of 4D CT in the range of seconds is possible. However, for events shorter than a second, a detector frame rate of thousands of frames per second (fps) and a highly intense X-ray source would be necessary, adhering to standard tomography protocol. This requirement surpasses the capabilities of current X-ray CT scanners. An exception to this is 4D CT monitoring of periodic events, where CT data acquisition can be synchronized with a suitable signal.

This study presents a novel triggered CT (tCT) system that delivers exceptional temporal and spatial resolution. Setting it apart from established methods is a unique CT data recording protocol where each projection is generated by integrating brief individual images captured at the same position of a periodically moving object. A critical requirement for this approach is the precise synchronization of all key elements of the tCT system, including the angular position of the CT scanner turntable and the X-ray detector read-out triggered by the accurate recording of the object's position. To ensure the precise and dependable acquisition of projections, the system employs real-time hardware synchronization of all components through specialized control software that interfaces with the X-ray detector data acquisition software.

The tCT system presented in this study has the potential for various applications, including three-dimensional modal analysis of oscillating objects and visualizing their internal structure. The oscillatory motion is facilitated by an electromagnetic shaker mounted on the CT rotary stage. For instance, a beam oscillating at 4 Hz can be tomographically captured during the amplitude of its periodic motion induced by the exciter, as shown in Fig. 1, which displays the reconstructed shape of the beam during oscillation. This approach allows for the identification of defects, such as component separation, that are challenging to distinguish statically. Additionally, it offers the opportunity to validate theoretical dynamic models that are difficult to verify through conventional methods.

Figure 1 compares the measurements of the same object using two different methods: the standard non-triggered CT (Fig. 1a) and the proposed triggered CT (tCT) method (Fig. 1b). In the standard untriggered CT image, the object is visible in all its states concurrently, with the amplitude positions being the most prominent, as the object statistically remains there the longest. However, due to the overlap of multiple positions, the outer envelope of the oscillating object is clearly visible, while the inner surfaces and object structure are blurred.

In contrast, the tCT method (Fig. 1b) provides a clear imaging of the entire oscillating object in its amplitude positions, which allows visualization of the actual shape of the oscillation and the internal structure of the object simultaneously. Both tomographies were acquired using a Dexela1512 detector and a Hamamatsu

L100321 X-ray source. In summary, the tCT method offers a significant improvement over the standard CT method without triggering and provides a clear and complete image of the oscillating object in the amplitude position.

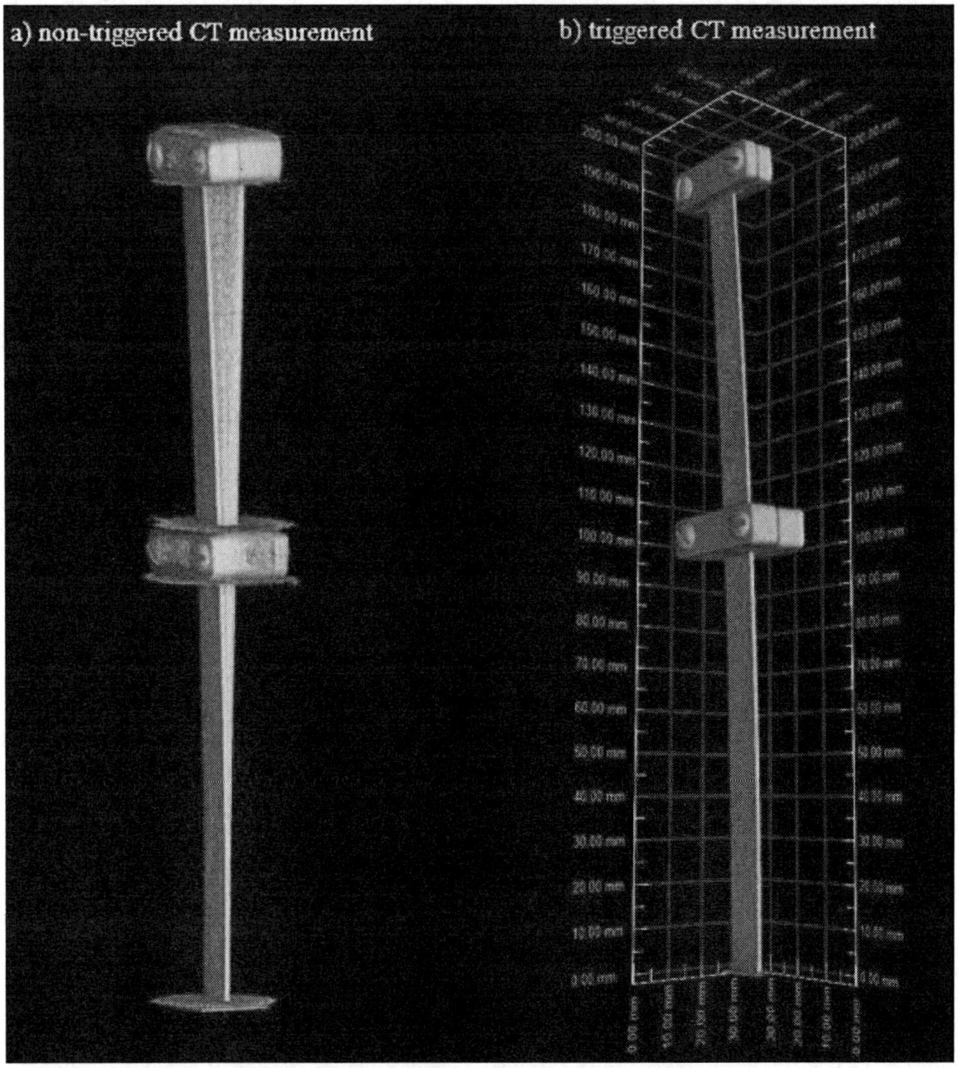

Fig. 1: Comparison of CT and tCT results.

Acknowledgement

This research has been financially supported by the Czech Science foundation (GA ČR) under project No. 22-13811S.

References

[1] D. Vavřík, T. Fíla, P. Koudelka, D. Kytýř, M. Macháček, V. Rada, J. Žemlička, P. Zlámal, X-ray computed tomography of the periodically moving object, Journal of Instrumentation 19 (2024), http://dx.doi.org/10.1088/1748-0221/19/01/c01045.

Bending Analysis of Glass Fiber Reinforced Composite Using Distributed Fiber Optic Sensing System

D. Blaha [1,a], Z. Padovec [1], K. Doubrava [1], R. Kottner [2], C. Novotný [1], N. Schmidová [1], M. Růžička [1]

[1] *Division of Strength and Elasticity, Fac. of Mech. Engineering, Czech Technical University in Prague; Technická 4, 166 07 Prague, Czech Republic*

[2] *ACO Industries Tábor s.r.o.; Průmyslová 1158, 391 01 Sezimovo Ústí, Česká republika*

[a] *David.Blaha@cvut.cz*

Abstract: In this study, we conducted a bending analysis of a glass fiber reinforced composite plate with a central hole, using a distributed fiber optic sensing system and other reference methods. The focus was to investigate the change in strain near the hole under bending stress. Our findings provide valuable insights into the behavior of such composites under prescribed load, contributing to their safer and more efficient use in various applications.

Keywords: distributed fiber optic sensing system; strain measurement; glass fiber composite; bending analysis.

1 Introduction

In recent years, water treatment solutions and technologies have been developed to address the growing demand for storing and treating water. One of the potential solutions is the use of glass fiber reinforced composited (GFRP) for their high strength to weight ratio. However, the behavior of GFRP under stress is complicated and requires detailed analysis to ensure their safe and efficient use especially in critical places such as near holes or other stress concentration points.

A distributed fiber optic sensing system (DFOS) is a powerful tool for measuring strain distribution along the length of an optical fiber. In comparison to conventional measurement methods, DFOS provides dense network of points for strain measurement. It is particularly useful when the strain distribution is not known in advance or when the strain distribution is expected to be non-uniform.

2 Methods

Strain measurements are performed using a distributed fiber optic sensing system (DFOS) and conventional strain gauges. These methods are used to measure the strain distribution along the hole of the composite plate. The DFOS system is based on the principle of Rayleigh scattering and is capable of measuring strain with high spatial resolution. The strain gauges are used as a reference method. The results from both methods are compared with FEM analysis.

2.1 Sensor Installation

The composite plate was made of glass fiber reinforced polyester resin. The sample is 400 mm long, 200 mm wide and 9.76 mm thick. The thickness corresponds to 19 layers of glass fiber. The central hole has a diameter of 80 mm.

The strain gauges were installed on the surface of the composite plate near the central hole in the direction of loading. The DFOS sensor was installed in parallel with the strain gauges, but on the opposite side of the hole. The DFOS sensor is a single-mode optical fiber installed in a wavy pattern on the surface next to the hole.

2.2 Testing

A setup for the four-point bending test is following. The distance between the upper supports is 120 mm and the distance between the lower supports is 320 mm. The strain gauges are installed on the upper (compression) side of the plate. The DFOS sensor is installed on both sides of the plate, one on the compression side and one on the tension side.

The testing was subjected to four-point bending on universal testing machine Heckert FPZ100/1. The samples were loaded with a ramp of 1 kN and then unloaded repeatedly. Then the samples were destructively loaded up to 10 kN.

3 Results and Conclusion

The results of the bending analysis of the GFRP composite plate with a central hole show that the strain distribution near the hole is non-uniform. The results from the DFOS system are in good agreement with the results from the strain gauges and FEM analysis. Loading up to 10 kN did not cause visible damage to the composite plate. Only residual strain was observed after unloading.

Acknowledgement

The project TN02000010 of National Competence Centre of Mechatronics and Smart Technologies for Mechanical Engineering is co-financed from the state budget by the Technology Agency of the Czech Republic within the National Centres of Competence Programme. Financial support was provided for this study by the Grant Agency of the Czech Technical University in Prague; grant no. SGS24/123/OHK2/3T/12.

References

[1] Luna Technologies, Optical Backscatter Reflectometer 4600, Technical report, 06 2006.

[2] C. Boller, F. Chang, Y. Fujino, Encyclopedia of structural health monitoring, Hoboken, NJ, Wiley-Blackwell, 2009.

[3] D. Munzke, E. Duffner, R. Eisermann, M. Schukar, A. Schoppa et al., Monitoring of type IV composite pressure vessels with multilayer fully integrated optical fiber based distributed strain sensing, in proc.: Materials Today (2021), Vol. 34, 217–223, https://doi.org/10.1016/j.matpr.2020.02.872.

[4] C. Karapanagiotis, M. Schukar, K. Krebber, Distributed fiber optic sensors for structural health monitoring of composite pressure vessels, tm – Technisches Messen 91 (2024) 168–179, https://doi.org/10.1515/teme-2023-0170.

[5] M. Froggatt, J. Moore, High-spatial-resolution distributed strain measurement in optical fiber with Rayleigh scatter, Applied Optics 37 (1998) 1735–1740.

Experimental Verification of the Influence of Thickness on 3D Printed Samples on Fracture Toughness Parameters

P. Bočan [1,a], A. Jíra [1]

[1] *Department of Mechanics, Faculty of Civil Engineering, Czech Technical University in Prague, 166 29 Prague, Czech Republic*

[a] *petr.bocan@fsv.cvut.cz*

Abstract: Currently, additive manufacturing technologies are widely used in biomedicinal applications. 3D printing technology brings great benefits to this industry as it can produce complex and diverse shapes. The problem then arises with very thin samples, where this technology is still not perfect and shows relatively large uncertainty and manufacturing defects. The main aspect of this research is to test the fracture toughness of very thin polyamide specimens fabricated by 3D printing using SLS technology. The fabrication of the specimens, the design and the experiment procedure are described in detail, followed by evaluation of the values and comparison with the values from the numerical model.

Keywords: fracture mechanics; 3D printing; polyamide PA12; digital image correlation; numerical analysis.

1 Introduction

In today's era of rapidly advancing technology and innovation, 3D printing has an indispensable place in manufacturing and scientific research across engineering disciplines. Material engineering and biomedicinal application are one of the key areas where 3D printing technology is being used. Titanium alloy 3D printing technology enables to creation of any shape of future implant, the design of an optimized surface and the fast individual manufacture of a specific replacement for a specific patient.

For the purposes of implants and their surfaces, it is necessary to create very thin structures with thicknesses ranging from 150 to 600 μm. This pushes the limits of current 3D printing capabilities, and the resulting mechanical properties do not align with theoretical expectations [1]. Current research often focuses on conventional mechanical parameters, but for obtaining a broader range of characteristics, it is necessary to expand experimental efforts [2]. This expansion is motivated by the subsequent development of a material model that corresponds to the reality of 3D printing. In clinical practice, metallic materials are primarily used for implants but are expensive to produce, and as this research is a pilot study it will investigate polymer samples (PA12 material) produced by 3D printing SLS technology. One of the main motivations for this research is to test fracture toughness, which is a key parameter for assessing the material resistance to accidental loading and crack propagation.

2 Experimental Analysis

First, I performed a pre-test experiment where I wanted to try out the testing methodology and then select the most suitable sample geometries for subsequent testing. In the next step of my research, I proposed to take into account the orientation of printing. The specimens were printed in horizontal (H) and vertical (V) directions. For both print orientations, 6 pieces were printed for each thickness (0.50 – 2.00 mm), giving a total of 140 printed specimens. This experiment involves a MARK 10 load press supplemented with a 3D DIC. In some cases, especially for thicker specimens, the crack has destroyed the specimen along its entire length. 3D DIC was used to determine the length of the crack a before the specimens were fully ruptured.

3 Numerical Analysis

A numerical model was created in GiD 16.0.6. for SAMPLE 2 and SAMPLE 3 with widths 0.50, 1.25 and 2.00 mm and the calculation was performed using ATENA software. Firstly, I determined the material model of PA12 material from data sheets and other experiments. A mesh with an element size of 0.25x0.25 mm was used in the most important region below the tip. For all 6 models, the calculation was performed with a linear and quadratic element mesh. The model was loaded with displacement and the Newton-Raphson method was used in calculation.

4 Conclusion

Table 1 shows the summary of fracture toughness values from the experiment calculated according to EN ISO 12 737. The results show a consistent increase for the horizontally printed samples from a certain width and very inconsistent result for thinner widths. The values for vertically printed samples are around 1.00 MPa $\sqrt{\text{m}}$ independent of the specimen width.

Tab. 1: Summary of experiment results of K_Q for PA12 material printed in horizontal and vertical directions with their deviation of results (H - horizontal print orientation, V - vertical print orientation) [3].

Thickness [mm]	0.50	0.75	1.00	1.25	1.50	2.00
SAMPLE 2 - H	2.0 ± 0.6	2.64 ± 1.12	1.65 ± 0.23	1.87 ± 0.09	2.16 ± 0.45	2.37 ± 0.31
SAMPLE 2 - V	0.99 ± 0.37	0.91 ± 0.38	0.93 ± 0.11	1.35 ± 0.35	0.94 ± 0.07	1.09 ± 0.15
SAMPLE 3 - H	1.54 ± 0.22	1.33 ± 0.48	1.5 ± 0.15	1.73 ± 0.38	2.01 ± 0.33	2.2 ± 0.36
SAMPLE 3 - V	1.2 ± 0.41	0.72 ± 0.11	1.3 ± 0.42	1.03 ± 0.06	1.21 ± 0.15	1.09 ± 0.15

Figure 1 shows a comparison of experimental data and numerical analysis for SAMPLE 2 with widths of 0.50 and 2.00 mm. Agreement between experimental and numerical analysis was not expected. One of the main reasons for discrepancies in results is the imperfect numerical material and geometric model that fails to account for imperfections in 3D printing (such as porosity) and the orthotropic properties of oriented 3D printing. The problem of 3D printing is that the technology is not yet good enough for the individual powder beads on the surface of the sample to melt together sufficiently. These powder beads can then "fall off" when the sample is handled, which weakens the thickness of the sample. This aspect is then fatal for very thin samples.

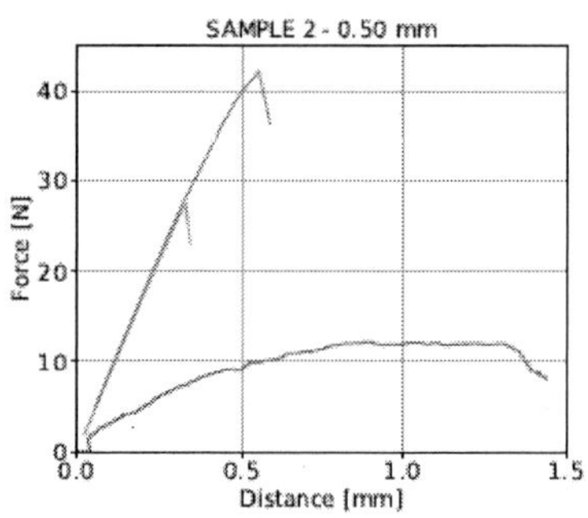

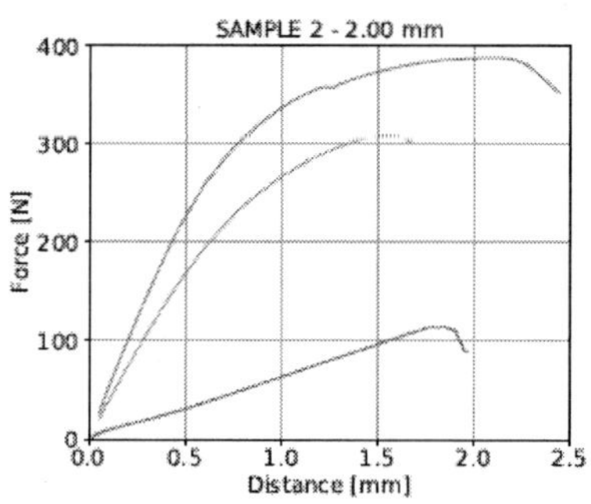

Fig. 1: L-D diagrams show comparison of values for SAMPLE 2 (0.50 mm and 2.00 mm) from the experiment and 2 types of numerical models with different mesh criteria (orange line - numerical model with linear elements, green line - numerical model with quadratic elements, blue line - experimental data) [3].

Acknowledgement

The financial support by Czech Technical University in Prague (SGS project No. SGS23/152/OHK1/3T/11) is gratefully acknowledged.

References

[1] A. Jíra et al., Mechanical properties of porous structures for dental implants: Experimental study and computational homogenization, Materials 14 (2021) 16, doi: 10.3390/ma14164592.

[2] P. Hájková, A. Jíra, Micromechanical analysis of complex structures by nanoindentation, in proc: 5th Annual Conference on Nanomaterials and Nanotechnology in Civil Engineering (2016), Prague, 60–65, doi: 10.4028/www.scientific.net/KEM.731.60.

[3] P. Bočan, Experimental verification of the influence of thickness on 3D printed samples on fracture toughness parameters, Diploma thesis, Prague, Czech Technical University in Prague, Faculty of Civil Engineering, Department of Mechanics, 2024.

Modeling the Compression Behavior of Additively Manufactured Lattice Structure and Energy Absorption

A. Boualleg [1,a], J. Fiori [2], D. Cirkl [1]

[1] *Technical University of Liberec, Czech Republic*
[2] *Universidade Estadual Paulista, Brazil*
[a] *abdelmadjid.boualleg@tul.cz*

Abstract: Lattice structures have garnered significant attention due to their lightweight nature and favorable mechanical properties. They find widespread application across various industries and engineering fields, such as automotive, aerospace, marine, and medical. Notably, the deformation of lattice structures allows for the dissipation of substantial amounts of energy. In this work, we need to understand the mechanical behavior of the lattice structure. In order to do that, we need to design the structure in the Software then choose the right flexible material for printing the model thereafter the model will be subjected to the experiments. The finite element method will be included for comparison of the experiment and simulation.

Keywords: lattice structure; TPU; viscohyperelastic; 3D printing.

1 Introduction

The lattice structures, characterized by their intricate network of interconnected struts, have gained prominence across various engineering domains due to their lightweight nature and unique mechanical properties [1]. These structures often manufactured through advanced techniques like additive manufacturing (AM), present an innovative approach to design challenges. The lattice geometry, with its high surface area and low material volume, offers an excellent compromise between structural integrity and weight efficiency [2].

This summary provides insights into different studies exploring lattice structures across diverse applications. From energy absorption and blast resistance to thermal energy isolation and ballistic performance, lattice structures exhibit versatile characteristics. The studies delve into optimizing lattice design, addressing manufacturing challenges, and enhancing specific properties crucial for various industries. Each investigation sheds light on specific aspects, contributing to the broader understanding and utilization of lattice structures in modern engineering.

2 Materials and Methods

2.1 Design

The lattice structure sample was designed in Abaqus Software with 10x10x9,6 mm cells with cylinder stems of 2 mm diameters. The cube has 6x6x5 cells in respective directions, giving 60x60x48 mm cube dimensions. This cube was printed in a 3D printer in an elastic material. For the material analysis, it was to design 3 samples in bone shape to do the tensile tests.

2.2 Experiment

The material used for printing the lattice structure is TPU; to investigate the mechanical properties of the material tensile tests were performed on the three bone shape samples.

The lattice structure was subjected to the compression test; the applied velocity is 0.05 mm/s velocity with stain 0.5. The experiment was compared with a simulation of the same structure in Marc Mentat 2022.4

Fig. 1: Lattice structure.

3 Conclusion

In conclusion, the experiment was conducted on lattice structures under uniaxial compressive force using both experiment and simulation. The TPU material used for printing the samples exhibits characteristics of viscoelasticity with damage. A comparison between the simulation and experimental data revealed a good in force-displacement relationships. Overall, the study highlights the importance of considering material properties and energy absorbed by the structure, which are increasingly being utilized in various engineering applications for their lightweight and efficient design.

Acknowledgment

This publication was written at the Technical University of Liberec as part of the project "Research of advanced materials, and application of machine learning in the area of control and modeling of mechanical systems" nr. SGS-2022-5072 with the support of the Specific University Research Grant, as provided by the Ministry of Education, Youth and Sports of the Czech Republic in the year 2022.

References

[1] J. León-Becerra, O. A. González-Estrada, J. Quiroga, Effect of Relative Density in In-Plane Mechanical Properties of Common 3D-Printed Polylactic Acid Lattice Structures, ACS Omega, 6 (2021) 29830–29838, doi: 10.1021/acsomega.1c04295.

[2] N. Thirunavukkarasu, J. Gao, S. Peng, A. Laroui, L. Wu, Z. Weng, Mechanically robust 3D printed elastomeric lattices inspired by strong and tough hierarchical structures, Additive Manufacturing, 66 (2023), doi: 10.1016/j.addma.2023.103451.

Various Applications of the Digital Image Correlation Method in the Experimental Analysis of Biomechanical Problems

G. Cabaj [1,a], P. Boháč [1], T. Návrat [1], P. Vosynek [1], V. Apostolopoulos [2], L. Kunz [3], J. Beneš [4]

[1] *Institute of Solid Mechanics, Mechatronics and Biomechanics, Faculty of Mechanical Engineering, Brno University of Technology, Technická 2896, 616 69 Brno, Czech Republic*
[2] *First Department of Orthopaedic Surgery, St. Anne's University Hospital and Faculty of Medicine, Masaryk University, Brno, Czech Republic*
[3] *Institute of Physics of Materials, Czech Academy of Sciences, Brno, Czech Republic*
[4] *BT Medical Engineering, Inc, Rozsochy, Czech Republic*
[a] *gabriel.cabaj@vutbr.cz*

Abstract: Nowadays, optical strain measurement using the digital image correlation (DIC) method is a frequently used experimental analysis method in various engineering fields. The main advantage is that it is a non-invasive method, therefore it is possible to analyze areas on structures that could be inaccessible to conventional contact methods. This paper aims to present various applications of the use of the digital image correlation method in biomechanical applications and specific case studies are presented using a commercial software Alpha from X-Sight company.

Keywords: biomechanics; strain analysis; digital image correlation; optical measurement; experimental modelling.

1 Introduction

Several experimental analyses using the DIC in the field of biomechanics are presented in this paper. It presents the possibilities of optical strain analysis, which compared to traditional contact methods, offers wider possibilities for experimental analysis of components. Examples are given where it is necessary to analyze the displacements of individual points on a component but also to analyze the whole region and evaluate the field of the strains. The purpose is to show the capability of the optical method, which comes where traditional contact methods of measuring deformations are not sufficient.

The experiments presented involve the determination of the strain distribution on an orthopaedic orthosis during a cyclic test. Later, the analysis of the composite pelvic bone model before and after the application of the fixator is described. The last two sections are dedicated to the analysis of the fixed fracture of the humerus and the compression test of implanted total knee endoprostheses in the composite tibial bone.

2 Strain Analysis of 3D Printed Orthopedic Orthoses

Supramalleolar orthosis is a type of orthotic device that is primarily used to stabilize the ankle and foot of patients. These orthoses are designed to provide control over the ankle joint but still allow for natural movement and sufficient flexibility. It is 3D printing technology that is the appropriate way to produce such devices. Thanks to its high flexibility, this method can be used to create personalized orthoses directly tailored to patients. In addition to the fact that 3D printing enables rapid prototyping, 3D printed materials also can exhibit much better mechanical properties than conventional materials. This significantly increases the durability and reliability of these orthoses [1–3].

The main objective of this work was the experimental comparative analysis of the mechanical response of the ankle orthosis to the loads to which it is subjected during the period of use by the patient. The ankle orthosis is primarily cyclically loaded daily, while the patient walks. However, significant loading occurs mainly when the orthosis is placed on the patient's foot. In this case, there must be a significant deformation of the orthosis for it to be placed on the foot [4]. The analysis of the strain field under quasi-static loading is primarily used to identify potential critical areas and can be used for component optimization. In this work, in addition to quasi-static loading, the response of the component to cyclic loading was also investigated.

3 Analysis of Strain and Displacement on a Composite Model of the Pelvic Bone

When the pelvic bone is injured and damaged, especially in the area of the symphysis, in common practice, it is then surgically fixed by the addition of a fixator. The problem that was analyzed in this case can be divided into two parts. In the first case, the effect of the addition of the fixator on the distribution of the stress in the pelvis compared to the undamaged pelvis was analyzed. This is particularly important from the point of view of the patient's recovery, to answer the question of whether a fixed pelvis would be more susceptible to damage outside the region of the fixator. In the second case, an analysis of the stress on the fixator after its application is required. This is because of its optimal design with regard to the risk of damage, especially by fatigue fracture. What is also analyzed is the effect of fixation - the number of screws [5].

The experiment was carried out on an electrodynamic INSTRON E10000 testing machine. The axis of the loading force was the hip joint and the junction of the pelvis with the sacrum. The whole experiment consisted of several load conditions. The response to quasi-static loading of the pelvis, with different magnitudes of loading force, was analyzed. The magnitude of the force was chosen with the consideration that the patient in recovery may start loading the pelvis gradually after surgery. Thus, each of these loading levels was evaluated. In the next phase, the force with a different amplitude from the quasi-static test was again applied, but the loading in this case was done cyclically with a frequency of 0.5 Hz and 1 Hz.

4 Analysis of Displacements after Fixation of Humerus Fracture

In a fracture of the humerus between the bone diaphysis and epiphysis, it is subsequently fixed with a fixator that is inserted into the axis of the bone. In this case, the aim was to determine how the humerus behaves with a fixed fracture. Specifically, whether there is significant relative motion between the humerus head and the diaphysis of the humerus. Subsequently, the influence of the type of fixation was also analyzed.

The humerus was loaded with a certain force in a universal testing machine with the help of a special fixture that simulated the 90° abduction that occurs when raising the arm. Thus, 500 cycles were performed and the motion of certain points was analyzed. In this case, DIC was only used to analyze the displacements of individual points from which the relative motion of the two parts, the humerus head and the diaphysis, could be determined.

5 Analysis of the Tibial Bone in the Compression Test

The final treatment of gonarthrosis involves the implantation of a total knee endoprosthesis that interacts with the tibial bone. The experimental analysis using DIC aimed to assess the effect of different types of implants on the mechanical response of the cortical bone tissue model to the composite tibial bone. This experiment was also part of a wider research with computational modelling using the finite element method [6–8].

The tibia with implanted tibial part of total knee replacement was loaded using a femoral component in compression. Two different total knee prostheses were analyzed in this way [6–8].

6 Conclusion

In this paper, several experimental analyses in the field of biomechanics using the digital image correlation method have been described.

This paper aims to present the wide possibilities of using the DIC system. The analyses described here would be difficult to perform using conventional contact methods.

References

[1] D. Al-Bayati, A.M. Kadhim, The Application of Supra_Malleolar Orthosis (SMO) in Iraq: Design and Fabrication Approach, IOP Conference Series: Materials Science and Engineering 1094 (2021) 012045, https://doi.org/10.1088/1757-899X/1094/1/012045.

[2] D. Popescu, F. Baciu, D. Vlăsceanu, R. Marinescu, D. Lăptoiu, Investigations on the Fatigue Behavior of 3D-Printed and Thermoformed Polylactic Acid Wrist–Hand Orthoses, Polymers 15 (2023) 2737, https://doi.org/10.3390/polym15122737.

[3] R. Kumar, S.K. Sarangi, 3D-Printed Orthosis: A Review on Design Process and Material Selection for Fused Deposition Modeling Process, in proc: Advances in Materials Processing and Manufacturing Applications (2021), A. Patnaik, E. Kozeschnik, V. Kukshal (Eds.), Springer Singapore, Singapore, 531–538, https://doi.org/10.1007/978-981-16-0909-1_55.

[4] M. Alvarez-Camacho, V. Alvarez-Guevara, C. Galvan Duque Gastelum, D. Flores Vazquez, G. Rodriguez-Reyes, F.M. Sanchez-Arevalo, Use of three-dimensional digital image correlation to evaluate mechanical response of prosthetic systems, Prosthetics & Orthotics International 41 (2017) 101–105, https://doi.org/10.1177/0309364616637956.

[5] J. Beneš, L. Kunz, J. Liška, J. Šrám, M. Krbec, V. Džupa, M. Křivohlávek, T. Návrat, Determination of stress and displacement on human composite pelvis under static and dynamic loading, Proceedings of the *Institution of Mechanical* Engineers, Part H 236 (2022) 951–959, https://doi.org/10.1177/09544119221099805.

[6] V. Apostolopoulos, Biomechanické a klinické srovnání totální náhrady kolenního kloubu typu All-poly a Metal-backed Online, disertační práce, Brno, Masarykova univerzita, Lékařská fakulta, 2024, dostupné z: https://is.muni.cz/th/le1t7/. [cit. 2024-04-18].

[7] V. Apostolopoulos, T. Tomáš, P. Boháč, P. Marcián, M. Mahdal, T. Valoušek, P. Janíček, L. Nachtnebl, Biomechanical analysis of all-polyethylene total knee arthroplasty on periprosthetic tibia using the finite element method, Computer Methods and Programs in Biomedicine 220 (2022) 106834, https://doi.org/10.1016/j.cmpb.2022.106834.

[8] V. Apostolopoulos, P. Boháč, P. Marcián, L. Nachtnebl, M. Mahdal, L. Pazourek, T. Tomáš, Biomechanical comparison of all-polyethylene total knee replacement and its metal-backed equivalent on periprosthetic tibia using the finite element method, Journal of Orthopaedic *Surgery* and Research 19 (2024) 153, https://doi.org/10.1186/s13018-024-04631-0.

Determining the Properties of GRP Tank Composite Materials

K. Doubrava [1,a], R. Kottner [2], Z. Padovec [1], R. Sedláček [1], M. Růžička [1]

[1] *Czech Technical University, FME; Technická 4, Praha, Czech Republic*
[2] *ACO Industries Tábor; Průmyslová 1158, Sezimovo Ústí, Czech Republic*
[a] *Karel.Doubrava@fs.cvut.cz*

Abstract: The paper describes the determination of the properties of composite materials and based on the results, the selection of the optimal choice for GRP tanks. The adhesive joints are used to connect the cylindrical part and the domes of the large capacity tanks, therefore various technologies for the preparation of adhesive joints were experimentally verified. The use of the sandwich construction of the wall of the vessel was investigated when the properties of the samples produced by the production technology were compared. The material properties of the core were investigated to create a numerical model of the tank.

Keywords: GRP tank; adhesive joints; sandwich; 3PB test; Double Cantilever Beam test; End Notched Flexture test.

1 Introduction

Surface water management is one of the main topics nowadays. The Glass Reinforced Plastic (GRP) surface water treatment tank is one of the important parts of the system chain solution. Optimising the design of the tank with regard to the price and technological complexity of production is one of the aspects in the search for new structural solutions for the resulting products. The load on the tank is determined both by the operational use, but also by the load created during the installation of the tank.

The connection between the dome and the cylindrical part of the tank is made using an adhesive joint. The optimal combination of technological surface preparation and suitable vinyl ester adhesives was investigated. The bending properties of the samples taken from the cylindrical part of the tank were investigated when the existing wall construction was compared with a sandwich version of the wall.

Fig. 1: Setup of Double Cantilever Beam test.

2 Methods

2.1 Properties of Adhesive Joints

Adhesive joints made from two variants of vinyl ester based bonding paste and two technological procedures for preparing the surface of the adherents were investigated. The evaluation of the variants was done on the basis [1] when the Double Cantilever Beam (DCB) test and the End Notched Flexture test (ENF) were described. The DCB test was based on the ASTM D5528 standard [2], when opening displacement was measured by optical extensometr, see Fig. 1. The ENF test (Fig. 2) was based on the ASTM D7905 standard [3], when maximal load point vs displacement was evaluated.

Fig. 2: Setup of End Notched Flexture test.

2.2 Bending Stiffness

Increasing the load capacity was one of the reasons for modifying the tank wall. The 3 point bending test of real samples was selected for the assessment of individual variants. Tank with different wall variants were produced and test specimens (Fig. 3) were taken from this tank. The bending stiffness of the samples (Fig. 4) was used to compare the variants. The compression test [4] and shear test of the core were done to obtain the material properties that were used in the numerical simulation.

Fig. 3: Bending specimens.

Fig. 4: Setup of 3PB test.

Conclusion

The results of the experiments were used to select the optimal combination of adhesive paste and production technology. Material data and test results of larger samples were then used to tune the numerical model of the entire tank, which is used to assess various structural modifications of the final product.

Acknowledgement

The work was supported by the project TN02000010 of National Competence Centre of Mechatronics and Smart Technologies for Mechanical Engineering is co-financed from the state budget by the Technology Agency of the Czech Republic within the National Centres of Competence Programme.

References

[1] R. Kottner, R. Hynek, T. Kroupa, Identification of parameters of cohesive elements for modeling of adhesively bonded joints of epoxy composites, Applied and Computational Mechanics 7(2) (2013) 137–144.

[2] ASTM D5528/D5528M-21, Standard Test Method for Mode I Interlaminar Fracture Toughness of Unidirectional Fiber-Reinforced Polymer Matrix Composites, ASTM International, 2022.

[3] ASTM D7905/D7905M-19e1, Standard Test Method for Determination of the Mode II Interlaminar Fracture Toughness of Unidirectional Fiber-Reinforced Polymer Matrix Composites, ASTM International, 2019.

[4] ASTM C365/C365M-22, Standard Test Method for Flatwise Compressive Properties of Sandwich Cores, ASTM International, 2022.

Flexural Behaviour of a Glass-Fibre Reinforced Polymer Composite: Experiment and Simulation

J. Ezenwankwo [1,a], I. Petrikova [1], J. Zak [1]

[1] *Technical University of Liberec, Liberec, Czech Republic*
[a] *Johnson.Ezenwankwo@tul.cz*

Abstract: This study involves the investigation of the flexural behaviour of a laminate composite. The three-point bending method is used with reference to the ASTM D7264/7264M – 07. The laminate is composed of a single ply with 20 layers of glass fibre as reinforcement. A stacking sequence of $[0_{20}]$ were used. Seven samples cut out from the entire board at 20x200 mm was used. The Thickness of the ply was 4.45 mm with a support span of 71mm according to standard. The samples were tested until visible delamination and debonding typified by visible shear and dip in applied compression up to a value of 5.8 mm. The simulation was done using Ansys ACP-pre and the results showed good agreement with the experimental result.

Keywords: glass fibre-reinforced polymer composite; flexural behaviour; experiment, simulation.

1 Introduction

Laminated composites have become a material of choice in modern day applications of engineering and technology. This attributed to its high strength-to-weight ratio, low machining cost and possibilities for extended studies in research and development. By studying the mechanical and thermal properties of composites, engineers, technologists and scientists are suitably equipped with the right experimental, numerical and simulation data to predict, with reasonable accuracy, the behaviour of materials in real-life applications. Be it advanced manufacturing techniques, natural-synthetic fibre-reinforced polymer composites, biocomposites, the application and scientific use of composites have grown in leaps and bounds, especially in the use of advanced techniques in investigating failures in composite materials [1, 2]. For years now, applications have forayed into automotive, aerospace, wind energy industries, and new frontiers have unfolded in smart composites and nanocomposites. Although fibre-reinforced composites have their tremendous benefits, they also have their inherent drawbacks and challenges arising from the fibre itself and the finished composite at various stages of their life cycle. Problems can arise in different stages including fabrication, joining, shaping, and disposal. Other concerns arise from functionality and usage, and include cost, (arising from high energy cost of production machining), quality assurance and control, durability, design complications and complexities, moisture conditions, fire resistance, compatibility with other materials, and adherence to applicable international standards for best practices [3–5]. The use of FE methods have created new ideas in modelling the real life behaviour and failure modes of composites – be it damage initiation, debonding and delamination of composite materials under bending load [6]. The critical natures of composite materials in use make it of utmost importance that they are prequalified and tested by standards to ensure they are, not only, fit for purpose but also have the highest level of safety and structural integrity. Nowadays there a number of high quality articles that highlight the effect of stacking sequence on the behaviour of composite materials in failure [7]. Improvements in the design of carbon fibre-reinforced polymer (CFRP) have led to the suitability of applications at elevated temperatures [8] and also highly dynamic loads in terms of multiaxial and fatigue stresses.

Fig. 1: Experimental instrumentation.

References

[1] Q. Shen, M. Omar, S. Dongri, Ultrasonic NDE Techniques for Impact Damage Inspection on CFRP Laminates, Journal of Materials Science Research 1(1) (2012) 34–36, DOI: 10.5539/jmsr.v1n1p2.

[2] S. Thabet, Y. Jasim, T. Thabit, Critically Evaluate the Capabilities of Ultrasonic Techniques Used for Tracing Defects in Laminated Composite Materials, International Journal of Engineering and Applied Sciences 10(3) (2018) 237–251, http://dx.doi.org/10.24107/ijeas.465749.

[3] R. N. Shama, T. G. A. Simha, K. P. Rao, G. V. V. Ravi Kumar, Carbon Composites are Becoming Competitive and Cost Effective, Infosys: NYSE, 2018.

[4] M. G. Bader, Selection of Composite materials and Manufacturing Routed for Cost-Effective Performance, Composites Part A: Applied Science and Manufacturing 33(7) (2002) 913–934, https://doi.org/10.1016/S1359-835X(02)00044-1.

[5] C. Lester, S. Nutt, Composite Materials: Advantages and Cost Factors, PhD, Material Science, 2018.

[6] H. Ullah, A. R. Harland , T. Lucas, D. Price, V. V. Silberschmidt, Computational Materials Science 52 (2012) 147–156.

[7] A. Azzam, W. Li., An Experimental Investigation on the Three-point Bending Behavior of Composite Laminate, IOP Conference Series: Materials Science and Engineering 62 (2014) 012016.

[8] R. M. Abdalrahman, Flexural Behaviour of CFRP Laminate at Elevated Temperature, Kurdistan Journal of Applied Research 2(3) (2017) 330–334, https://doi.org/10.24017/science.2017.3.26.

Impact of Direct Ageing on Mechanical Properties and Residual Stresses of Maraging 300 Steel Manufactured by SLM

R. Halama [1,a], M. Kořínek [1], J. Čapek [2], K. Trojan [2], J. Hajnyš [1]

[1] *VŠB-Technical University of Ostrava, 17. listopadu 2172/15, Ostrava-Poruba, Czech Republic*
[2] *Czech Technical University in Prague; Trojanova 13, 120 00 Prague, Czech Republic*
[a] *radim.halama@vsb.cz*

Abstract: This paper describes the main results of research conducted on samples manufactured by selective laser melting of maraging steel 300 powder. The results of the tensile tests helped us to construct maps of yield and ultimate strength as a function of temperature and time hold of direct ageing of the material. Furthermore, X-ray diffraction was used to determine the decrease of residual stresses with time during heat treatment under the lowest temperature. The magnitude of residual stresses in the axial direction of the sample was reduced by 60 %.

Keywords: additive manufacturing; residual stresses; X-ray diffraction; maraging steel; heat treatment.

1 Introduction

Selective laser melting (SLM) is a technology that falls into the group of additive manufacturing (AM). SLM is a promising technique for consolidating metal powder and presents a plethora of opportunities for the creation of parts. Using a scanning energy source entails selectively melting sections of a thin, flat powder bed layer by layer (20–100 μm thick) to create 3D parts [1]. Complex residual stresses (RS) distribution is produced during the SLM process when regions are heated in accordance with the scan method, allowing parts to heat and cool independently. The application of heat treatment (HT) prior to removing the part from the build plate and removing supports is an efficient approach to reduce RS in printed parts. To prevent microstructural alterations, the temperature for HTs must be carefully chosen; otherwise, the high strength of the material may be lost.

Investigating the impact of HT on mechanical characteristics and residual stresses of maraging 300 steel (1.2709 or MS300) is the objective of this work. MS300 is a nickel-iron alloy with good strength, exceptional toughness, and good weldability. The HT is usually based on precipitation hardening. Steel is used extensively in the automobile, aerospace and tooling industries [2].

2 Material and Methods

The yield and maximum strength of MS300 increase during HT mainly due to the precipitation of the intermetallic nanoparticles $Ni_3(Mo, Ti)$ and Fe_2Mo [3]. The impact of direct ageing at various temperatures and different holding times on the mechanical properties of MS300 samples prepared with SLM has been reported in a preliminary study [4]. In this study, we present new results from tensile tests considering HT: 490 °C, 520 °C, 840 °C, 900 °C (always 1, 2, 6 and 10 hours). All tensile tests were performed using a LabControl 100 kN / 1000 Nm hydraulic testing machine. Residual stress X-ray diffraction measurements were done on samples subjected to HT at 490 °C.

3 Results

There are two local extrema in the map of yield strength evaluated from tensile tests on heat treated samples, see Fig. 1a. The temperature of 490 °C was selected for the investigation of averaged RS measured by X-ray diffraction around the circumference of MS300 in the axial direction as a function of hold time (Fig. 1b).

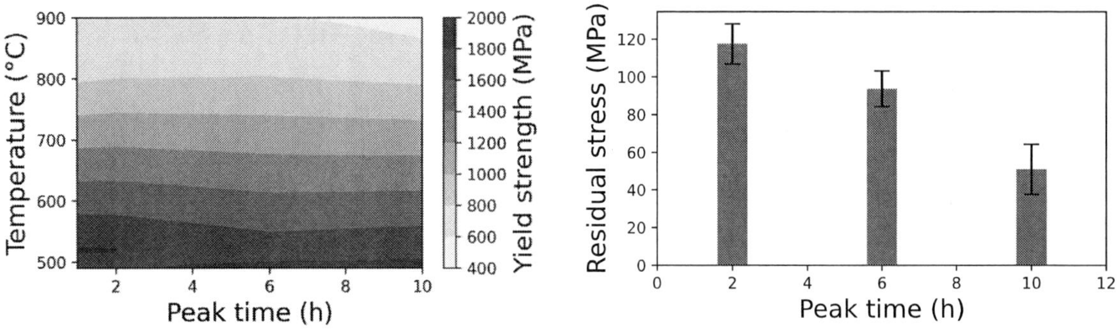

a) A map of the yield strength resulting from HTs b) RS magnitude evolution with hold time

Fig. 1: Results of mechanical tests and X-ray measurements on MS300 SLM.

4 Conclusion

In total, sixteen HT conditions, as a combination of four temperatures and four time holds, have been researched. The best case from the strength increase point of view has been selected for the evaluation of the decrease in RS due to HT. The magnitude of the axial component of residual stresses decreased to 40% of its initial value. The study findings indicate that RS cannot be completely eliminated by 10 hours of the HT.

Acknowledgement

This work is an output of the project with financial support from the Czech Science Foundation under the registration no. 23-04724S. The work of CTU staff was supported by the Grant Agency of the Czech Technical University in Prague, grant No. SGS22/183/OHK4/3T/14.

References

[1] R. Halama et al., Cyclic plasticity of additively manufactured metals, In book: Cyclic Plasticity of Metals: Modeling Fundamentals and applications, Amsterdam, Elsevier, 2022, 397–433, https://doi.org/10.1016/B978-0-12-819293-1.00022-X.

[2] E.-A. Jagle et al., Precipitation and austenite reversion behavior of a maraging steel produced by selective laser melting. Journal of Materials Research 29 (2014) 2072–2079, https://doi.org/10.1557/jmr.2014.204.

[3] T. Allam et al., Tailoring the nanostructure of laser powder bed fusion additively manufactured maraging steel, Additive Manufacturing 36 (2020) 101561, https://doi.org/10.1016/j.addma.2020.101561.

[4] A. Kvita, Research into the Influence of Heat Treatment on the Mechanical Properties of the MS300 Material Produced by the SLM Method, Diploma thesis, Ostrava, Faculty of Mechanical Engineering, VŠB-TU Ostrava, 2022. (in Czech)

Self-Heating Phenomenon of Magnetorheological Elastomer MRE under Dynamic Loading

M. Hdaib [1,a], I. Petrikova [1]

[1] *Technical University of Liberec, Czech Republic*
[a] *Mohammad.yousef.hdaib@tul.cz*

Abstract: In this study, a finite element analysis (FEA) of the self-heating phenomenon inside MRE under compressive dynamic loading is presented based on the large-strain viscoelasticity model. The material behavior was investigated experimentally using quasi-static uniaxial compression and frequency sweep dynamic compression. Then the least-squares method and FEA were used to estimate and validate the model parameters. After that, the self-heating phenomenon was simulated using the FEA and compared with experimental data. According to the FEA of the quasi-static compression, the static friction on the sample can be neglected. Moreover, the fully coupled FEA predicts the self-heating phenomenon slightly better than the uncoupled one. Finally, the self-heating phenomenon can be predicted accurately using the uncoupled FEA with information about exposed surfaces' temperatures and hysteresis loop at the steady-state conditions unless the material has a phase change or high stress softening due to temp.

Keywords: MRE; material model; heat build-up; thermal analysis; composite material.

1 Introduction

The rapid development in Industrial and engineering applications requires adaptable and suitable new types of composite materials. Smart materials are responsive composite materials with controllable mechanical properties that can be altered by external stimuli such as magnetic field, temperature, or light. Like conventional materials, smart materials could fail due to mechanical or thermal fatigue. Therefore, it is essential to investigate the self-heating phenomenon to point out possible applications and failure modes of smart materials. Magnetorheological elastomers (MREs) are an important type of smart materials capable of adjusting their mechanical properties and shape in the presence of a magnetic field. MRE basically consists of an elastomer matrix filled with iron particles [1]. They are suitable for applications that require variable stiffness like soft robotics, vibration isolators, and sensors.

In this study, a fully-coupled FEA of the self-heating phenomena was performed at variable frequency. A large strain viscoelasticity model was used with a generalized Maxwell model. A Quasi-static uniaxial compression test was employed to fit the hyperelastic parameters. In contrast, a frequency sweep uniaxial compression was performed to estimate the viscoelastic parameters in the frequency domain. The frequency sweep has a span from 10 Hz to 50 Hz and a step size of 2 Hz. Then the heat generation results were discussed and compared with the previous experiment performed by the authors [2].

2 Methodology

The hyperelastic and viscoelastic behavior of the material was investigated using quasi-static and dynamic compression, respectively. The relations shown in Eq. (1) and Eq. (2) were used to estimate the material constants. Where σ is the engineering stress, while E_S and E_L are the storage and loss modulus, respectively.

$$\sigma = 2C_{10}\left(\lambda - \frac{1}{\lambda^2}\right) + 2C_{01}\left(1 - \frac{1}{\lambda^3}\right) \tag{1}$$

$$E_S = E_\infty + \sum_{i=1}^{3}\left(\beta_i \frac{\omega^2\tau_i^2}{1+\omega^2\tau_i^2}\right) \quad , \quad E_L = E_\infty \sum_{i=1}^{3}\left(\beta_i \frac{\omega^2\tau_i^2}{1+\omega^2\tau_i^2}\right) \tag{2}$$

Consequently, FEA was applied using the acquired parameters to simulate the self-heating phenomenon. The FE mesh and thermal boundary conditions are illustrated in Fig. 1 and Tab. 1 illustrate the thermal properties of the MRE samples.

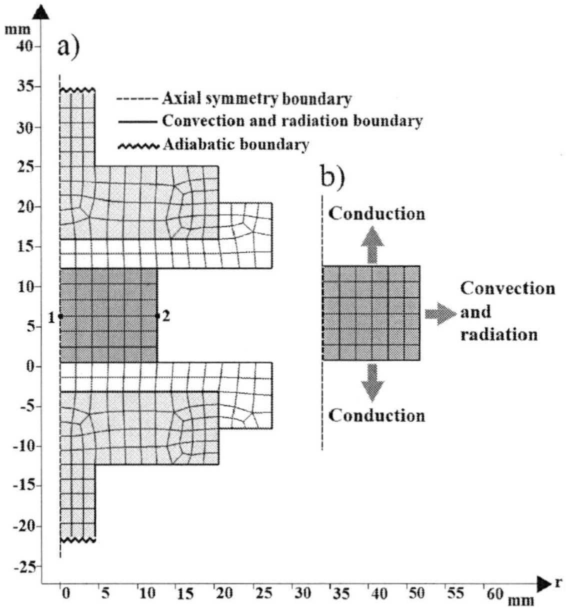

Fig. 1: FE mesh and thermal boundary conditions.

Tab. 1: Thermal properties of the MRE sample, Teflon attachments, and steel grips [2].

	MRE	Steel	Teflon
Emissivity [-]	0.95	0.22	0.85
Conductivity [w.m^{-1}.K^{-1}]	0.775	44.5	0.24
Specific heat [J.kg^{-1}.K^{-1}]	585.08	475	1050

3 Conclusion

The simulated temperatures in the center and the sample surface (points 1 and 2 in Fig. 1) showed satisfactory with experimental data. Based on our observation, fitting the Prony-series in the frequency domain is better to simulate higher frequency applications; when the frequency is higher than 20 Hz.

Acknowledgement

This work was supported by the Student Grant Competition of the Technical University of Liberec under project No. SGS-2022-5072.

References

[1] A.K. Bastola, M. Hossain, The shape – morphing performance of magnetoactive soft materials performance, Materials & Design 211 (2021) 110172, https://doi.org/10.1016/j.matdes.2021.110172.

[2] T.H. Nam, I. Petríková, B. Marvalová, M.Y. Hdaib, Self-heating and dynamic mechanical behavior of silicone rubber composite filled with carbonyl iron particles under cyclic compressive loading, Journal of Composite Materials 55 (2021) 4273–4292, https://doi.org/10.1177/00219983211037055.

Load Capacity Analysis of Sheathed Core Static Rope

P. Hisem [1,a], J. Kolář [1,b], J. Kult [1,c]

[1] *Department of Virtual and Applied Mechanics, VÚTS a.s., Svárovská 619, 460 01 Liberec XI-Růžodol I, Czech Republic*

[a] *pavel.hisem@vuts.cz,* [b] *jan.kolar@vuts.cz,* [c] *jaroslav.kult@vuts.cz*

Abstract: The paper describes and evaluates the tensile strength testing of the polyamide sheathed core static rope and its components. The paper also describes and evaluates rope tensile tests to determine the influence of knots and the influence of the interaction of the rope and rounded edges of various radii on the tensile strength.

Keywords: static rope; tensile test; load capacity; polyamide; sheathed core; rope knot.

1 Introduction

The tested static rope meets the requirements of the EN 1891, which defines the properties of static ropes with sheathed cores. The rope testing methodology is determined by the standard EN 364, in justified cases the test methods deviated from this standard. The goal of the work was not to confirm the already well-known basic properties of the rope but to better understand the mechanics of the rope.

2 Experiments

The static polyamide rope diameter of 10.5 mm was investigated. The load capacity of the entire rope, rope sheath, rope core, simple-knot rope, knot-ended rope, and rope in interaction with the rounded edge of various diameters was tested (see Fig. 1).

a) entire rope test

b) rope sheath test

c) rope core test

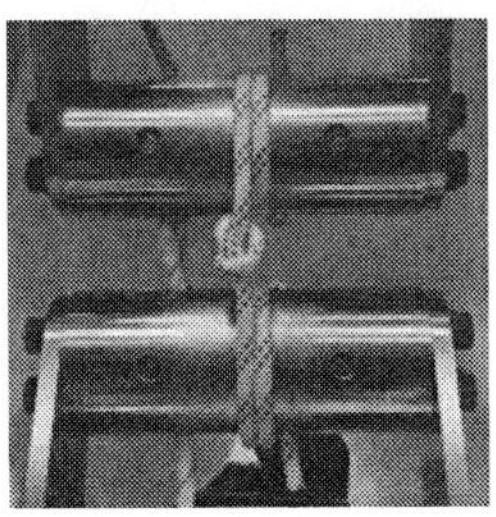

d) simple-knot rope test

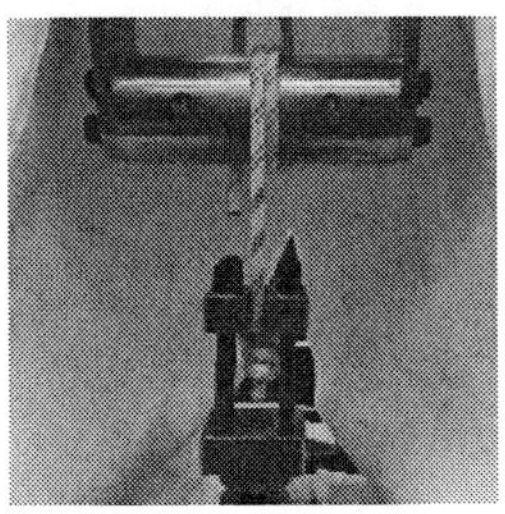

e) knot-ended rope test

f) rounded edge test

Fig. 1: Tensile test of rope, rope sheath, rope core, rope with knots and in interaction with the rounded edge.

3 Results

Results of load capacity tests of static rope are shown in the following graphs (see Figs. 2, 3, 4).

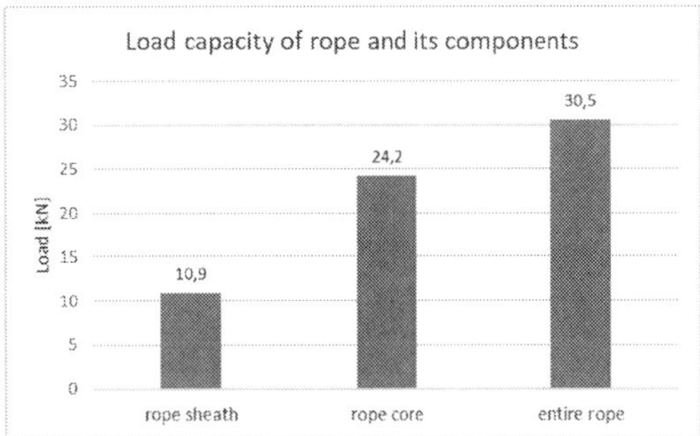

Fig. 2: Test results of rope and its components.

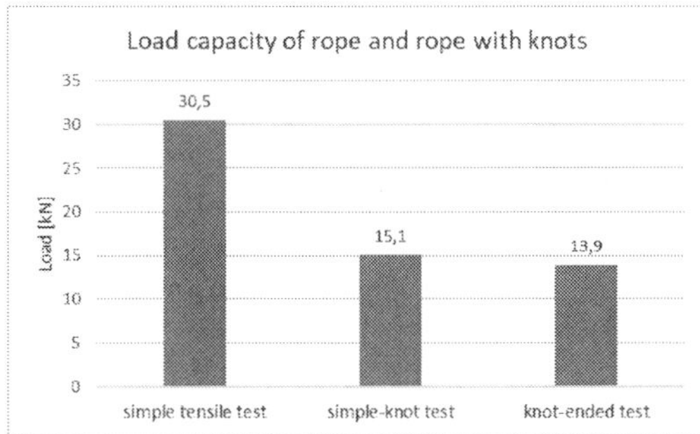

Fig. 3: Test results of rope and rope with knots.

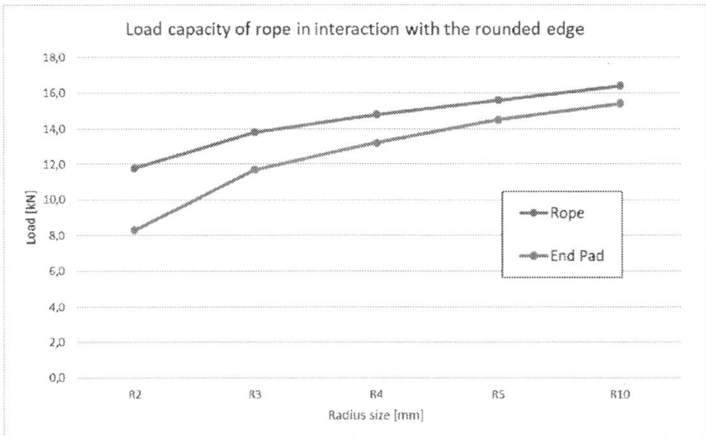

Fig. 4: Results of load capacity test of rope in interaction with the rounded edge of various diameters.

4 Conclusion

The sum of the load capacity values of the rope components (sheath and core) is approx. 15% higher than the measured value of the entire rope. The load capacity of the rope in a simple tensile test was measured 30.5 kN. If we consider this value as a base (100%), then the measured load capacity of a rope in a simple-knot test of 15.1 kN corresponds to a relative reduction to 50% and the measured load capacity of a rope knot-ended test of 13.9 kN corresponds to a relative reduction to 46%. The load capacity test of rope in interaction with the rounded edge of various radii carried out confirmed the assumption about the negative effect of small radii on the load capacity, when radius R2 has a relative reduction to 39% of the simple tensile loaded rope.

Acknowledgement

This publication was supported by the Ministry of Industry and Trade (MPO) within the framework of institutional support for long-term strategic development of the research organization – provider MPO, recipient VÚTS, a.s.

References

[1] ČSN EN 1891 (2000) Personal protective equipment for the prevention of falls from a height. Low stretch kernmantel ropes.

[2] ČSN EN 364 (1996) Personal protective equipment against falls from a height – Test methods.

Methodology and Determination of Parameters for Modeling Brittle Adhesive

L. Horák [1,a], J. Krystek [1]

[1] *Department of Mechanics, Faculty of Applied Sciences, University of West Bohemia, Univerzitní 8, 301 00 Pilsen, Czech Republic*

[a] *horakl@kme.zcu.cz*

Abstract: In the numerical simulation of adhesive joints, a significant challenge lies in obtaining reliable experimental data due to the wide variety of adhesives available, each with distinct material properties and different material models. This study focused on testing the methodology for obtaining mechanical parameters using bulk adhesive specimens. The experiments were conducted on a brittle adhesive, specifically Araldit AV 138+HY 998, which was selected as a representative material for structural adhesives. The study also evaluated fracture parameters, including the strain energy release rate, an essential parameter for predicting the fracture behavior of adhesive joints. By focusing on a specific adhesive material, this study aimed to obtain its mechanical parameters and use them in numerical simulations.

Keywords: adhesive; mechanical testing; bulk specimen; DCB; ENF.

1 Introduction

Mechanical testing of adhesives is crucial in ensuring bonded assemblies' reliability and durability. Bulk specimens are commonly used for this purpose, as they provide a convenient and standardized way to measure the mechanical properties of adhesives.

Bulk specimens are typically prepared by casting or molding the adhesive into a standardized shape, such as a dumbbell or cylinder. The specimen is then subjected to various mechanical tests, including tensile, compression, and shear tests. These tests measure the stress-strain behavior of the adhesive, which is used to calculate important mechanical properties such as the modulus of elasticity, yield strength, and ultimate tensile strength. [1]

One of the main advantages of using bulk specimens for mechanical testing is that they allow for measuring the adhesive's properties independent of the substrate material. This is important because the substrate material can influence the mechanical properties of the adhesive, and testing the adhesive alone provides a more accurate measure of its intrinsic properties.

However, it is essential to note that the mechanical properties of bulk specimens may not always correlate directly with the performance of the adhesive in real-world applications. This is because the mechanical behavior of adhesives can be influenced by various factors, including the surface preparation of the substrate, the thickness and geometry of the bond line, and the environmental conditions during application and use. [2]

To address these challenges, researchers have developed a range of advanced testing methods that can more accurately simulate the conditions experienced by adhesives in real-world applications. These methods include dynamic mechanical analysis, fracture mechanics testing, and environmental testing.

All these parameters are then applicable to associated material models in numerical simulations, which, with suitable criteria for adhesive failure, can determine safe loads and the duration of these loads. [3]

2 Experiments

A series of experiments were conducted to obtain the mechanical parameters specified in Table 1, and associated samples were manufactured. After several iterations, a thin plate of adhesive was cast in a simple mold and then placed under a vacuum to ensure uniformity and remove any air bubbles. Following the hardening period, specific samples were milled to have defined geometry and thickness, as required for the various experiments.

Tab. 1: Parameters for complex numerical simulation.

Zone name	Parameter name	Symbol	Unit	Experiment
Elastic	Young modulus in tension	E	[Pa]	Tensile test
	Poisson constant	ν	[-]	Tensile test
	Young modulus in shear	G	[Pa]	Arcan test
Iniciation of Fracture	Tensile stress limit	σ_{max}	[Pa]	Tensile test
	Shear stress limit	σ_{max}	[Pa]	Arcan test
	Ultimate tensile deformation	ϵ_{max}	[-]	Tensile test
Crack propagation	Strain energy release rate for mod I	G_I	[Jm^{-2}]	DCB
	Strain energy release rate for mod II	G_{II}	[Jm^{-2}]	ENF

Several experiments were conducted for the elastic zone parameters, including Young's modulus in tension, Poisson's constant, and Young's modulus in shear, including tensile tests, extensometer measurements, and Arcan tests. These tests were performed to determine the material's elastic properties, which are crucial for predicting the behavior of adhesive joints under various loads.

The initiation of fracture parameters, such as tensile stress limit, shear stress limit, and ultimate tensile deformation, were evaluated using tensile and Arcan tests. These parameters are essential for predicting the point at which the adhesive joint begins to fail.

For the crack propagation parameters, including the strain energy release rate for modes I and II, samples for Double Cantilever Beam (DCB) and End-Notched Flexure (ENF) tests were fabricated. These tests were conducted to evaluate the fracture behavior of the adhesive joint under different loading conditions.

Since the adhesive tested in this study is brittle and elastic, no mechanical parameters for plasticity were assumed. The experiments focused on determining the elastic, fracture initiation, and crack propagation parameters, which are critical for predicting the behavior of adhesive joints under various loads and conditions.

3 Conclusion

This work summarizes the methodology for obtaining complex mechanical parameters for brittle adhesives and measures these parameters for adhesive Araldit AV 138+HY 998.

Acknowledgement

This research work was supported by the project SGS-2022-008.

References

[1] A.M.G. Pinto et al., Shear Modulus and Strength of an Acrylic Adhesive by the Notched Plate Shear Method (Arcan) and the Thick Adherend Shear Test (TAST), Materials Science Forum 636–637 (2010) 787–792, doi: 10.4028/www.scientific.net/MSF.636-637.787.

[2] M.D. Banea, L.F.M da Silva, Adhesively bonded joints in composite materials: An overview, Proceedings of the Institution of Mechanical Engineers, Part L: Journal of Materials: Design and Applications 223 (2009) 1–18, doi: 10.1243/14644207JMDA219.

[3] G. Dean, B. Duncan, Preparation and Testing of Bulk Specimens of Adhesives, Teddington, Middlesex, National Physical Laboratory, 1998, ISSN 1368-6550.

Enhanced Detection of Rotor-Stator Contacts Using Continuous Wavelet Transform

V. Houdek [1,a], L. Smolík [1], Z. Kubín [1]

[1] *Research and Testing Institute Plzen, Tylova 1581/46, 301 00 Plzeň, Czech Republic*
[a] *houdek@vzuplzen.cz*

Abstract: This research presents an advanced method for detecting rotor-stator contacts in large rotating systems such as turbines and compressors in power plants. The method combines continuous wavelet transform (CWT) with signal decomposition techniques to improve the sensitivity and reliability of contact detection. The proposed method is validated using signals measured during the operation of a 1000 MW turbine generator. The results show significant improvements in the early detection of rotor-stator contacts over standard methods based on the short-time Fourier transform, contributing to better maintenance practices and extended equipment life.

Keywords: rotor-stator contact detection; continuous wavelet transform; vibration analysis; signal noise reduction; predictive maintenance.

1 Introduction

Large rotors used in supercritical power plants or other highly effective turbomachinery are susceptible to rotor-stator contacts due to relatively low clearances, which can lead to severe mechanical failures if not detected timely. This research introduces an advanced method employing the continuous wavelet transform (CWT) combined with denoising techniques based on signal decomposition, which can detect such contacts earlier and more reliably than traditional methods based on the short-time Fourier transform (STFT).

2 Methodology

The study utilizes a standard condition monitoring system to obtain raw data. Typically, rotor vibrations are measured at bearing stations using velocity seismic transducers and inside bearing pedestals using eddy-current proximity sensors. The latter can capture the relative motion of the shaft to the bearing, which incorporates the dynamic interaction between the rotor and stator. The continuous wavelet transform (CWT) is employed to analyze the recorded vibration signals, enabling the identification of transient behaviours indicative of rotor-stator contacts.

If the signal is a sequence of numbers, perhaps samples of some function of a continuous variable or a set of inner products, the expansion of that signal is called a discrete-time wavelet transform (DWT). It maps a sequence of numbers into a sequence of numbers much the same way the discrete Fourier transform (DFT) does. Since the recorded signals are composed of discrete samples (with a constant sampling rate of 2048 Hz), the DWT is used to estimate the CWT of the signal. The DWT employs Haar wavelet, a sequence of square-shaped functions that allows the best localization of features in the transformed signal in time. Conversely, any feature would be poorly localized in frequency compared to continuous or smooth wavelets. The DFT is sequentially employing a cascade of lowpass and highpass filters. The resulting decomposed signals represent the behavior of the original signal in varying frequency ranges. The decomposed signals are called $d1$ to $d5$ coefficients for clarity, with $d1$ representing the frequency range with the highest frequencies.

3 Innovations in Detection Techniques

Introducing denoising processes using the 3σ rule and soft thresholding significantly enhances the signal's clarity by reducing noise interference, a common challenge in dynamic system monitoring. This method improves the detection reliability of rotor-stator contacts under varying operational conditions.

4 Experimental Validation

Data recorded during the operation of a 1000 MW turbine generator were analyzed, with particular attention to transient behaviors during rotor-stator contacts. Four distinct datasets corresponding to different operational scenarios were used to test the effectiveness of the proposed method. Fig. 1 shows some illustrative results.

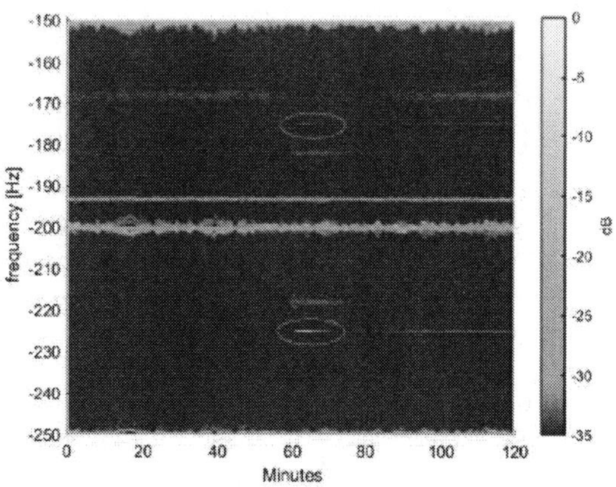

a) full spectrum analysis zoomed (-250) - (-150) Hz - rub symptoms in ellipses; -3.5X (-175Hz) and -4.5X (-225Hz) sub-harmonics

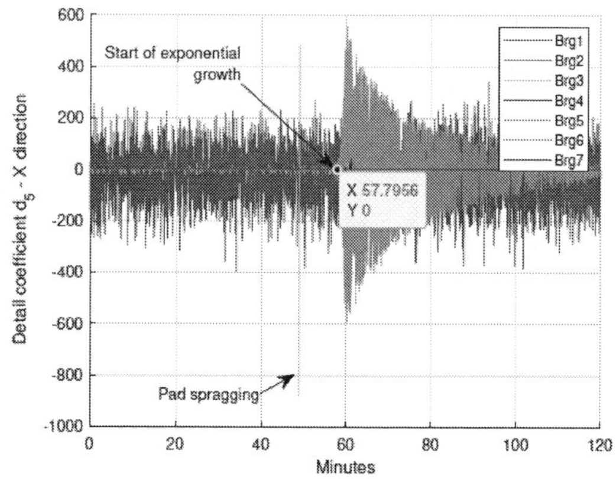

b) detail coefficients d_5 for bearings 1 to 7; $d_5$at BRG5 raised at 57.8 minute

Fig. 1: Dataset 2 standard evaluation process.

5 Results

The enhanced detection method successfully identified rotor-stator contacts in multiple test scenarios, demonstrating superior sensitivity and specificity compared to traditional FFT-based methods. The ability to detect subtle changes in vibration signals before they manifest into larger issues was notably improved.

6 Conclusion

The proposed method, based on signal filtering and subsequent continuous wavelet transform, represents a significant step forward in the non-destructive diagnostics of large rotating machinery, with potential applications extending beyond the power plant applications to aerospace, automotive, and other industries reliant on heavy rotational equipment. Traditional methods utilize the STFT to obtain asynchronous vibration, which can be a symptom of a rotor-stator contact. However, the signal-to-noise ratio (SNR) of the STFT can be unsatisfactory, which leads to less accurate localization of the symptoms in time. The proposed method improves the SNR almost by the order of magnitude, providing much better results. Moreover, the method can be optimized for each transducer, further improving the SNR. The validation indicates that the method is not sensitive to vibration changes due to operating parameters, which renders the method attractive for early warning systems or anomaly detection systems that operators can use for online condition monitoring purposes.

Acknowledgement

This work was supported by the institutional support for long-term conceptual development of research institutions provided by the Ministry of Industry and Trade of the Czech Republic.

Determination of the Dynamic Bending and Torsional Stiffness of a Beam using Enhanced FRF

R. Huňady [1,a], P. Palička [1], E. Dubňanská [1], M. Hančinová [1]

[1] *Technical University of Košice, Faculty of Mechanical Engineering, Letná 1/9, Košice, Slovakia*
[a] *robert.hunady@tuke.sk*

Abstract: This paper deals with the estimation of dynamic stiffness coefficients from the so-called enhanced frequency response functions (EFRFs), which are extracted from the FRF matrix obtained in the modal test. The given procedure is presented on the example of a free straight beam with constant cross section.

Keywords: bending stiffness; torsional stiffness; estimation; modal analysis; EFRF.

1 Introduction

The enhanced frequency response function is based on the concept of transforming physical coordinates to modal coordinates, which makes it possible to isolate a selected mode and express it in terms of the response of a system with one degree of freedom. The enhanced frequency response function of *r*-th mode can be determined as follows [1]

$$EFRF(\omega)_r = \{U(\omega_r)\}^H\,[H(\omega)]\,\{V(\omega_r)\}\,, \tag{1}$$

where $\{U(\omega_r)\}$ is a mode shape vector of *r*-th mode, $\{V(\omega_r)\}$ is a modal participation factors vector of *r*-th mode, obtained by singular value decomposition of the FRF matrix $[H(\omega)]$ done for each spectral line

$$[H(\mathrm{i}\omega)] = [U(\omega)]^H\,[S(\omega)][V(\omega)]\,, \tag{2}$$

where $[S(\omega)]$ is a diagonal singular value matrix. The plot of singular values corresponds to the Complex Mode Indicator Function

$$CMIF(\omega)_k = S_k(\omega)\,. \tag{3}$$

Once the EFRFs for the first bending and first torsional mode are extracted, then the stiffness coefficient is determined from the stiffness line $\kappa(\omega)$ tangent to the left part of the EFRF curve. Depending on the type of FRF, the stiffness is given by

$$\kappa(\omega) = \frac{1}{k}\ \text{- for receptance,}\quad \kappa(\omega) = \frac{\mathrm{i}\omega}{k}\ \text{- for mobility,}\quad \kappa(\omega) = \frac{-\omega^2}{k}\ \text{- for inertance.} \tag{4}$$

2 Experimental Modal Analysis

The object of the modal test was a steel beam of rectangular cross-section with dimensions 40x8x400 mm. The beam was loaded quasi-freely using two rubber bands. A B&K impact hammer Type 8206 was used for excitation. The responses were measured using a Polytec PDV100 laser vibrometer. The geometric model (Fig. 1) consisted of 18 input DOFs and 1 output DOF. After the measurements, the frequency response functions were converted from mobility to receptance and then processed according to the procedure described in Section 1. The first bending and the first torsional mode shape of the beam are shown in Fig. 2.

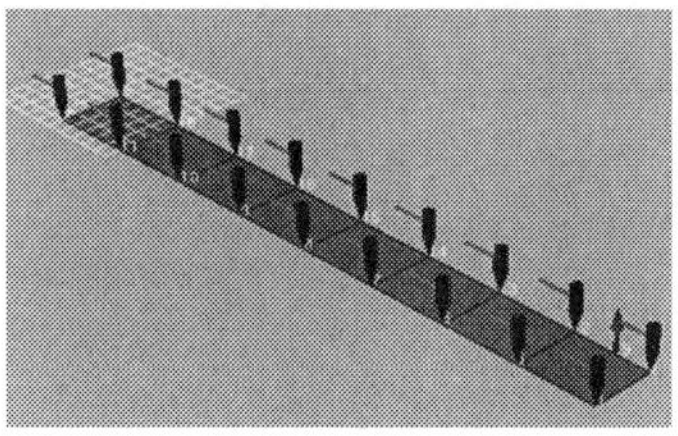

Fig. 1: Geometrical model.

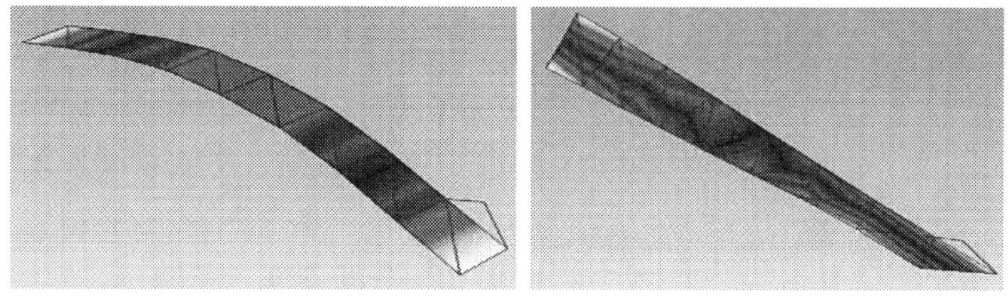

Fig. 2: The 1. bending (266.5 Hz) and the 1. torsional mode (1496.8 Hz).

3 Estimation of Stiffness Coefficients

Figure 3 and Fig. 4 show the EFRF plots of the first bending and the first torsional mode, respectively, including the CMIF plot, stiffness line, mass line and synthesised EFRF. Since the spectra shown represent the receptance, i.e. the mechanical compliance, the stiffness coefficient is given as the inverse value corresponding to the stiffness line. For the bending mode, the bending stiffness coefficient can be determined directly in this way, i.e. $k_b = \frac{1}{\kappa(0)}$, but the torsional stiffness coefficient must be recalculated as $k_t = \frac{\kappa(0)\cdot b^2}{4}$, where b is the beam width. The resultant coefficients are listed in Tab. 1.

Tab. 1: Coefficients of dynamic bending and torsional stiffness of a free beam.

Mode	$\kappa(0)$	k_b (N/m)	Mode	$\kappa(0)$	k_t (Nm/rad)
1. bending	$1.9532\cdot10^{-6}$	511 980	1. torsional	$1.19424\cdot10^{-7}$	3349

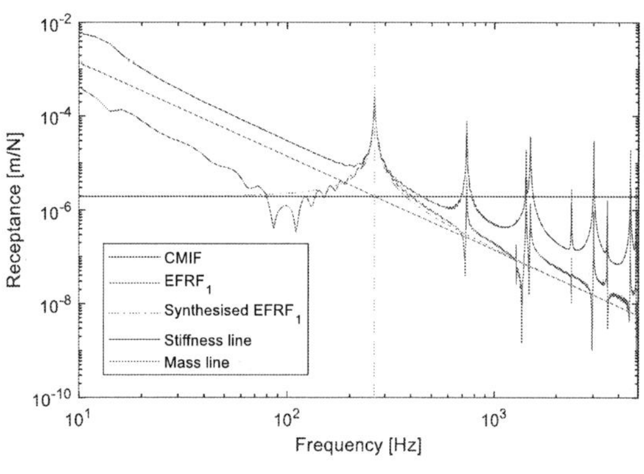

Fig. 3: EFRF of the first bending mode.

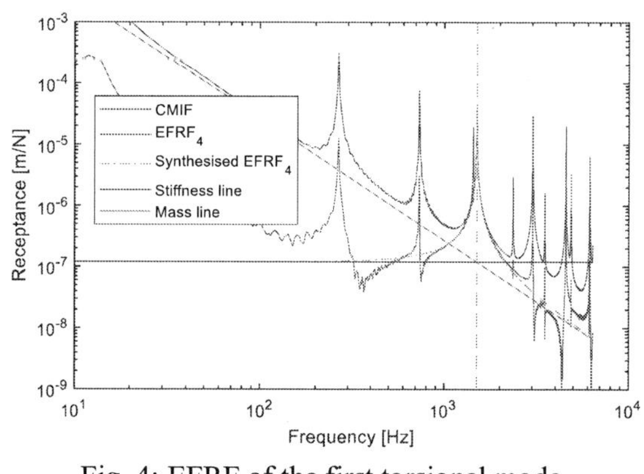

Fig. 4: EFRF of the first torsional mode.

Acknowledgement

This work was supported by the Science Grant Agency of the Ministry of Education, Science, Research and Sport of the Slovak Republic and the Slovak Academy of Sciences under the project VEGA 1/0500/20.

References

[1] R.J. Allemang, D.L. Brown, A complete review of the complex mode indicator function (CMIF) with applications, in proc: International Conference on Noise and Vibration Engineering (ISMA) (2006), Katholieke Universiteit Leuven, Belgium, 3209–3246.

Experimental Study of Mechanical Properties of Titanium Alloy with Beta Structure

A. Jíra [1,a], Z. Tolde [2], J. Fojt [3]

[1] *Czech Technical University in Prague, Faculty of Civil Engineering, Department of Mechanics, Czech Republic*

[2] *Czech Technical University in Prague, Faculty of Mechanical Engineering, Department of Physics, Czech Republic*

[3] *University of Chemistry and Technology Prague, Faculty of Chemical Technology, Department of Metals and Corrosion Engineering, Czech Republic*

[a] *jira@fsv.cvut.cz*

Abstract: 3D printing of beta-titanium alloys for biomedical applications is currently in high demand, both for material reasons and for the ability to produce highly complex replacements, often directly tailored to the patient. These individual implants require a porous surface that is printed from thin parts at the limits of the 3D printer's capabilities. The thin structures are then very susceptible to changes in mechanical properties due to poor print quality, and building a corresponding numerical model is very complicated. The aim of this work is to experimentally describe the basic mechanical properties of thin 3D-printed titanium alloy samples with a beta structure. These values can then serve as input quantities for numerical modeling via particle models.

Keywords: 3D printing; surface etching; titanium alloy; porosity; mechanical testing.

1 Introduction

In the past decade, there has been a revolutionary transformation in manufacturing due to 3D printing, enabling the creation of intricate structures with high precision. In biomedicine, this technology has become pivotal for the development of implants and joint replacements that can better match the individual anatomical characteristics of patients [1, 2].

The key advantage of using 3D printing with these materials lies in the utilization of complex porous structures, such as gyroid-based designs [3], which ensure optimal integration of the implant with surrounding bone tissue. However, these structures are often very thin (typically 150 – 600 μm), pushing the limits of current printers and the powders used.

In real-world applications, experimental measurements often do not correspond with numerical models due to internal material porosity and incomplete fusion of powder particles at the structure's edges [4].

This work serves as a pilot study on the behavior of thin 3D printed samples, aiming to experimentally assess their mechanical properties. The values obtained will form the basis for creating a numerical model. Numerical modeling and simulations are crucial for the precise design and optimization of these structures based on desired mechanical properties.

2 Methodology

Sample preparation was focused on 3D printing (type Dog bone) specimens for uniaxial tensile testing to obtain fundamental material characteristics for subsequent numerical modeling. For this purpose, both conventional titanium alloy Ti6Al4V and an experimental alloy Ti25Nb4Ta8Sn were utilized. The Ti6Al4V specimens underwent surface etching, while the Ti25Nb4Ta8Sn specimens were evaluated in different groups:

- Group I: Ti25Nb4Ta8Sn samples without supplementary post-processing treatments.
- Group II: Ti25Nb4Ta8Sn samples subjected to post-annealing to alleviate internal stresses.
- Group III: Ti6Al4V specimens without additional post-processing treatments.
- Group IV: Ti6Al4V samples subjected to surface etching and removal of imperfectly fused powder grains through washing.

These experimental groups were essential for evaluating and comparing the mechanical properties and structural integrity of the printed specimens under different conditions.

3 Results

Uniaxial tensile testing was conducted as a fundamental comparison to illustrate the development of mechanical properties in thin samples. The standard Ti6Al4V alloy exhibits significantly higher ultimate tensile strength values compared to the Ti25Nb4Ta8Sn alloy. This difference can be attributed to the significantly higher porosity in the beta-structured alloy. This high porosity (~16–19 %) also leads to a substantial reduction in strength after annealing, although there is not a notable change in the Young's modulus. Conversely, for the conventional alpha-beta alloy, there was no decrease in ultimate tensile strength for etched samples, but significant decreases in Young's modulus were observed for the thinnest samples. As the thickness increases, the differences in the force-displacement curves diminish, suggesting that the quality of 3D printing plays a less critical role.

Tab. 1: Average values of yield strength and ultimate tensile strength for different types of samples.

		Ti25Nb4Ta8Sn		**Ti6Al4V**	
t [μm]		with annealing	without annealing	with etching	without etching
500	F_{max} [N]	407.5 ± 11.8	207.7 ± 29.6	1009.4 ± 43.5	760.0 ± 85.4
	σ_{max} [MPa]	203.8 ± 5.9	135.3 ± 14.8	504.7 ± 21.8	380.3 ± 75.0
750	F_{max} [N]	889.6 ± 38.6	674.0 ± 86.1	3570.4 ± 129.5	3571.1 ± 105.8
	σ_{max} [MPa]	296.5 ± 12.9	224.7 ± 28.7	1190.1 ± 83.2	1190.4 ± 61.9
1000	F_{max} [N]	1442.6 ± 69.9	1010.9 ± 101.0	4579.4 ± 177.1	4958.9 ± 133.0
	σ_{max} [MPa]	360.6 ± 16.7	252.7 ± 25.3	1144.9.7 ± 144.9	1246.5 ± 45.7
		Group I	Group II	Group III	Group IV

Acknowledgement

We are thankful for the financial support provided by The Czech Science Foundation (GACR) grant No. 23-04971S and Czech Technical University in Prague grant SGS No. SGS23/152/OHK1/3T/11.

References

[1] D. Serrano et al., 3D Printing Technologies in Personalized Medicine, Nanomedicines, and Biopharmaceuticals, Pharmaceutics 15 (2023).

[2] B. Li et al., Application and Development of Modern 3D Printing Technology in the Field of Orthopedics, BioMed Research International 2022 (2022) 1–15.

[3] A. Jíra et al., Variability of the Gyroid Structure for Biomedical Applications, in proc: 59th International Scientific Conference on Experimental Stress Analysis (2021), Litomyšl, Czech Republic.

[4] A. Jíra et al., Mechanical Properties of Porous Structures for Dental Implants: Experimental Study and Computational Homogenization, Materials 14 (2021).

Measurement of Fatigue Loading of Bus Bodies and Components on Model and Real Tracks and Test Polygons

M. Kepka [1,a], J. Vaclavik [2]

[1] *University of West Bohemia, Univerzitni 8/2732, 301 00 Pilsen, Czech Republic*
[2] *Research and Testing Institute Plzen., Tylova 1581/46, 301 00 Pilsen, Czech Republic*
[a] *kepkam@fst.zcu.cz*

Abstract: The main types of measurement of fatigue loading of bus bodies and components are presented, which allow assessment of their fatigue life.

Keywords: buses; fatigue loading; model track; real track; test polygon; fatigue life.

1 Introduction

Operating buses, trolleybuses, and battery buses on uneven roads causes dynamic and variable amplitude loading of these road vehicles. Such loading can lead to fatigue damage to critical nodes of the bodywork or other vehicle components [1]. Time courses of service loads are most often random processes.

Such loading processes can only be determined with sufficient accuracy by measurements. Fatigue analysis of these processes is fundamentally different from dynamic analysis.

Dynamic analyses are performed in the frequency domain; fatigue analyses must be performed in the time domain to determine the so-called load spectra (histograms of load cycles). Fatigue damage can then be estimated using an appropriate fatigue damage accumulation hypothesis.

Knowing the maximum load values is sometimes sufficient for assessing structural nodes of the bodywork and the critical cross-sections of the components of the mentioned vehicles.

2 Load Cases

The measurement is usually performed with an empty bus and a load representing a fully occupied vehicle. For both of these load cases, the stresses in the critical bodywork nodes and component´s cross sections can be determined at various excitation types, see Fig. 1.

The most important approach is the measurement while the vehicle prototype is driving in real traffic on various unevennesses of real roads. The measured random stress-time histories are converted to fatigue damage using cumulative damage rules, which have been proposed by various authors, for example, Haibach [2]. For the assessed structural nodes, information about their fatigue properties is necessary. The S-N curves are evaluated based on the results of laboratory fatigue tests, or they are taken from standards.

Some customers also require measuring the vehicle during its tests on the proving ground containing special tracks with defined parameters of unevennesses. Based on such measurements, the potential for shortening the operational fatigue test on the test polygon can be estimated [3].

The load's maximum amplitudes (ranges) can be measured when the vehicle drives over an artificial obstacle of the given shape. The cylinder segments are laid on an even asphalt road in such a composition that overruns occur gradually by right-hand wheels, both wheels simultaneously and left-hand wheels. Such measurements make it possible to estimate the future service stress spectra of the assessed structural node [4].

Accelerations are also measured. The evaluated power spectral densities can then, under certain assumptions, be used for vibration fatigue calculations or vibration tests.

Measuring systems are also being developed for long-term online monitoring of operating loads of critical structural nodes and vehicle components.

Fig. 1: Types of forced body vibration: crossing artificial standardised obstacles, driving on real roads with different surface quality, and driving on special tracks on the test polygon [5].

3 Conclusion

The contribution describes the measurement of fatigue loading of bus bodies and components on model and real tracks and test polygons. Such measurements and presented practical examples are based on the authors' long-term cooperation with various bus manufacturers.

References

[1] M. Kepka, M. Kepka jr., Deterministic and probabilistic fatigue life calculations of a damaged welded joint in the construction of the trolleybus rear axle, Engineering Failure Analysis 93 (2018), https://doi.org/10.1016/j.engfailanal.2018.07.015.

[2] E. Haibach, Modified Linear Damage Accumulation Hypothesis Accounting for a Decreasing Fatigue Strength During the Increasing Fatigue Damage, TM Nr. 50. LBF, Darmstadt, Germany, 1970.

[3] M. Kepka, M. Kepka jr., Accelerated fatigue testing on special tracks as new part of methodology for bus/trolleybus development, Engineering Failure Analysis 118 (2020), https://doi.org/10.1016/j.engfailanal.2020.104786.

[4] M. Kepka, M. Kepka jr., Design, service and testing grounds stress spectra and their using to fatigue life assessment of bus bodyworks, in proc: 12th International Fatigue Congress (2018), MATEC Web of Conferences, 165, 17007.

[5] M. Kepka, M. Kepka jr., Forced Vibration of Bus Bodyworks and Estimates of Their Fatigue Damage, In Recent Trends in Wave Mechanics and Vibrations, in proc. WMVC (2023), Springer Conference.

Influence of Temperature Change on Young Modulus of Elasticity of CFRP

B. Kropík [1,a], M. Matušů [1], Z. Padovec [1], T. Mareš [1]

[1] *Department of Mechanics, Biomechanics and Mechatronics, Faculty of Mechanical Engineering, Czech Technical University in Prague, Technická 4, 160 00, Prague, Czech Republic*

[a] *bohumil.kropik@fs.cvut.cz*

Abstract: Repeated tensile testing at low stress levels was done at different temperature levels to estimate the influence of temperature on the Young modulus of elasticity of the composite material. Testing was done for temperatures higher than the room temperature and also for lower temperatures with the use of cooling by liquid nitrogen. Tests were held for two different types of fibres and the results indicate there is a change in the Young modulus of elasticity depending on the temperature.

Keywords: CFRP; Young modulus; elasticity; temperature.

1 Introduction

Composite materials, which in this context refer to Carbon Fibre Reinforced Polymers (CFRP), are often designed to enhance the mechanical strength of structures with the aim of reducing the assembly weight or increasing structural potential. Operating temperatures often do not differ significantly from room temperature, but in some cases they may vary considerably. Some information on temperature-related behaviour regarding the mechanical properties of CFRP can be found in [1] for temperatures higher than room temperature and in [2] for temperatures lower than 0 °C. Therefore, for our design purposes, it was decided to measure the relationship between the Young modulus of elasticity and temperature. CFRP used in this study comprised two types of carbon fibres: High Strength (HS), partially documented in [3], and Ultra-High Modulus (UHM) fibres, together with epoxy resin. These specimens were subjected to repeated measurements during the elastic phase of the tensile test to determine whether the temperature variations affect the stiffness of the composite material and to what extent, particularly in our case.

2 Overview of the Contribution

The specimens were designed with a focus on the usage of the same production method as the final product. Thus, the beam made by filament winding on the mandrel of square cross-sectional shape with thickness about 5 mm was cut into plates sized 1000 x 120 x 5 mm. The specimens were then cut from these plates to a size of 200 x 40 x 5 mm. Together, 10 specimens of HS composite and 10 specimens of UHM composite were prepared for testing. Each specimen was equipped with a linear 1-axis Strain Gauge (SG) installed on each side, orientated in the loading direction. Initially, each specimen was loaded at room temperature, followed by loading at various temperature levels above and below room temperature, and often rechecked at room temperature afterward with the use of our thermal chamber. The measurements made at decreased temperatures were done with the use of liquid nitrogen to cool the environment inside the chamber. The loading scenario was meant to not damage the specimens, so the specimens were loaded up to 10 kN of tensional force in displacement mode of 1 mm/min in three cycles on each thermal level and the evaluation of Young modulus of elasticity was between 4 and 5.5 kN using the SG signal.

3 Conclusion

An experimental determination of the Young modulus of elasticity was conducted at specific temperature levels. The results indicate a change in the Young modulus of elasticity with temperature variation, in which could be taken into account during design. Additionally, some specifics regarding the measurement of the specimens at reduced temperatures were identified. More tests are necessary, especially with liquid nitrogen cooling, to determine the proper design of the experimental setup and to identify the behaviour of the specimen.

Fig. 1: Specimen in a thermal chamber during loading.

Acknowledgement

This study received support from the Grant Agency of the Czech Technical University in Prague, under grant Nr. SGS24/123/OHK2/3T/12.

References

[1] Y. Yang et al., Study on Tensile Properties of CFRP Plates under Elevated Temperature Exposure, Materials 12 (2019), doi: 10.3390/ma12121995.

[2] Z. Jia et al., An experimental investigation of the temperature effect on the mechanics of carbon fiber reinforced polymer composites, Composites Science and Technology 154 (2017), doi: 10.1016/j.compscitech.2017.11.015.

[3] M. Matušů et al., Influence of Temperature Increase on Young Modulus of Elasticity of CFRP, in proc.: Applied Mechanics 2021: 22nd International Science Conference for Young Researchers, Book of Full Texts (2021), Technical University of Liberec, Liberec, 36–41.

Fatigue Life Prediction of 42CrMo4+QT Steel Using Self-heating Effect

M. Matušů [1,2,a], J. Papuga [1]

[1] *Department of Mechanics, Biomechanics and Mechatronics, Faculty of Mechanical Engineering, Czech Technical University in Prague, Technická 4, 160 00, Prague, Czech Republic*

[2] *Department of Mechanical and Environmental Engineering, OTH Amberg-Weiden, Germany*

[a] *martin.matusu@fs.cvut.cz*

Abstract: This study proposes a new approach that leverages the self-heating effect during cyclic loading, showing promise for the prediction of fatigue life. The self-heating phenomenon offers valuable insight into material behavior, but the stabilized temperature increase used generally as the key parameter might be absent in some cases. Introducing a modified version of Fargione's method, the study addresses such scenarios in the temperature response of specimens from 42CrMo4+QT steel examined here. Experimental investigations on five datasets, each with different machining parameters, analyze how they influence fatigue life and residual stresses. The findings indicate the proposed approach could be also applied to materials lacking temperature increase stabilization. Challenges related to variable limiting energy are explored, along with proposed solutions for future evaluation. In summary, this study introduces a promising path for assessing fatigue life in materials with an unconventional temperature response during cyclic loading.

Keywords: Fargione's method; self-heating effect; 42CrMo4; fatigue life; surface quality.

1 Introduction

Materials can fail under cyclical loading even when stress levels remain below their tensile or yield strength. The relationship between stress amplitude (S) and the number of cycles to failure (N) is represented by an S-N (stress-life) curve. The traditional experimental methods for fatigue life assessment are resource-intensive, requiring extensive testing and numerous specimens to establish the S-N curve [1]. Standard guidelines recommend 6-12 specimens for research and 12-24 for structural design, making reduction of the number of specimens highly desirable but without sacrificing accuracy. The fatigue life prediction methods explored here utilize surface temperature measurement during cyclic loading, which causes the temperature increase due to the self-heating (S-H) effect. Promising results were reached for steel [2] and aluminum [3] specimens.

This study introduces a refined approach to the prediction of fatigue life, based on Fargione [2], but with a focus on the S-H effect. Our investigation specifically addresses cases where temperature evolution lacks Phase 2 (see Fig. 1), as observed in previous research [4]. Five configurations were analyzed made of conventionally manufactured 42CrMo4+QT steel with varying surface finishes, yielding diverse fatigue life performances.

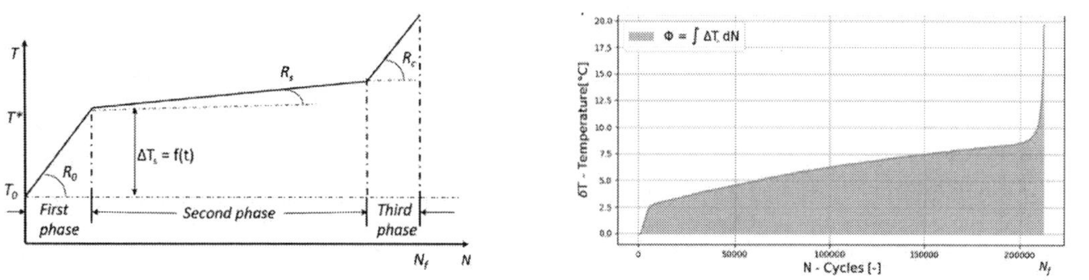

a) Schematic description of the S-H effect. b) The Self-Heating effect during the experiment.

Fig. 1: Schematic depiction of temperature response during constant-amplitude cyclic loading until failure N_f divided into three different phases, but without any stabilized region in Phase 2 and a real measured record.

2 Experimental Campaign

Five series of tests were performed on specimens from 42CrMo4+QT high-strength steel using an Amsler resonant pulsator, focusing on the fatigue strength. Each of the 6 configurations of machining setups resulted in varying residual stresses, leading to different fatigue strengths and fatigue limit (FL) transitions. Fatigue life was described using regression analysis, as illustrated in Fig. 2.

To predict fatigue life, a modification of Fargione's method was derived to enable the use of the non-stabilized temperature section, and to improve the agreement with the experimentally set S-N curves. Due to a limited number of specimens, series A25-A28 were tested with 10 specimens for the S-N curve and one for the S-H step-test, analyzing the self-heating effect at different amplitudes. In the case of the A29 series, additional specimens for the S-H test were available. From S-H tests, the prediction of the S-N curve was established as illustrated in Fig. 2.

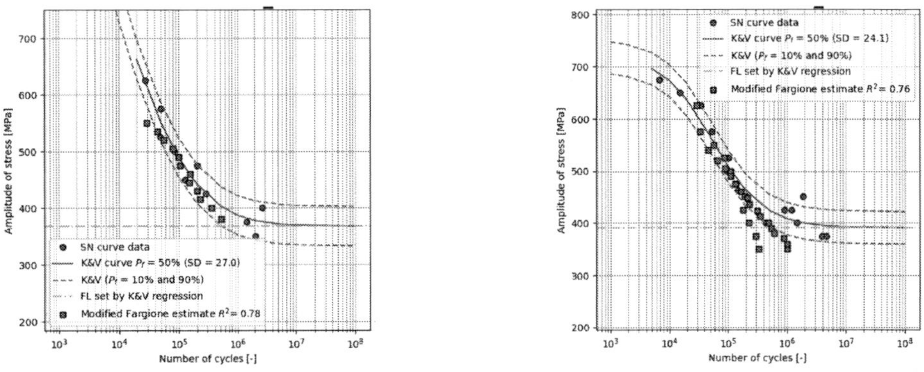

a) A26 series.

b) A29 series.

Fig. 2: Estimates of the S-N curves by the Modified Fargione's method for series A26 and A29.

3 Conclusion

This work introduces a novel approach for predicting fatigue life in 42CrMo4+QT steel, a material lacking a stabilization phase in its self-heating temperature evolution. By modifying Fargione's approach, the S-N curve could be estimated based on a single sample subjected to a Self-Heating step-test, yielding intriguing results.

Acknowledgement

The authors acknowledge the support by the Grant Agency of the CTU in Prague (SGS23/156/OHK2/3T/12 project) and by the Czech Ministry of Education, Youth and Sports (LUABA22071 project).

References

[1] Standard Practice for Statistical Analysis of Linear or Linearized Stress-Life (S-N) and Strain-Life (e-N) Fatigue Data 1981.

[2] G. Fargione, Rapid determination of the fatigue curve by the thermographic method, International Journal of Fatigue 24 (2001) 11–19, https://doi.org/10.1016/S0142-1123(01)00107-4.

[3] M. Matušů, K. Dimke, J. Šimota, J. Papuga, J. Rosenthal, V. Mára, L. Beránek, Energy-based method for analyzing fatigue properties of additively manufactured AlSi10Mg, Journal of Mechanical Science and Technology 37 (2023) 7, https://doi.org/10.1007/s12206-022-2110-6.

[4] J. Papuga, M. Mžourek, M. Matušů, V. Mára, J. Čapek, Investigation of the size effect on 42CrMo4 QT steel in the high-cycle fatigue domain part I: Experimental campaign, International Journal of Fatigue, 175 (2023) 107743, https://doi.org/10.1016/j.ijfatigue.2023.107743.

On a Possibility of High-Resolution Neutron Diffraction Observation of TiAl Substructure

P. Mikula [1,a], A. Michalcová [2], V. Ryukhtin [1]

[1] *Nuclear Physics Institute ASCR, 250 68 Řež, Czech Republic*
[2] *Department of Metals and Corrosion Engineering, University of Chemistry and Technology, 166 28 Prague 6, Czech Republic*
[a] *mikula@ujf.cas.cz*

Abstract: The feasibility of using an unconventional high-resolution neutron diffraction three-axis setting for possible observation of the substructure of TiAl alloy is presented. The usefulness of the diffraction setting is documented in the measurements of powder diffraction and small-angle neutron scattering profiles of several samples of TiAl alloy as prepared by a different technology. TiAl samples of $\phi = 9$ mm were made from the 3D printed material of gamma titanium aluminides (Ti48Al2Cr2Nb) loaded on the bulk Ti substrate. Though the measurement at the medium power research reactor is rather time-consuming, it is possible to obtain some complementary results related to studies of plastic deformation and porosity of the individual samples.

Keywords: neutron diffraction; small-angle neutron scattering; TiAl alloy; plastic deformation; experimental analysis.

1 Introduction

Thanks to their excellent properties (e.g. low density, superior strength, creep, and oxidation resistance), the γ-TiAl alloys are considered as promising materials fulfilling the demands for next generation high-temperature materials in the aerospace, automotive, energy industries etc. The material density of γ-TiAl alloys is approximately half of the conventional Ni-based superalloys, their use can significantly reduce the weight of e.g. engines [1]. Much attention is paid to the investigation of the influence of the technological preparation of the TiAl alloys on their properties, which naturally, are strongly dependent on the alloy structure. In our contribution we present the results of the feasibility studies of high-resolution neutron powder diffraction profiles of three samples of TiAl alloy in relation to their technological preparation. We demonstrate high-resolution properties of the 3-axis neutron diffraction setting permitting the finer studies of diffraction lines. The earlier and present obtained results have shown that the high-resolution setting could be used in some special studies exploiting the diffraction line profile analysis e.g. for micro-strain/micro-stress analysis of plastically deformed samples, determination of possible phases having very close values of the lattice spacings [2]. TiAl samples were made from the 3D printed material of γ-TiAl (Ti48Al2Cr2Nb) loaded on the bulk Ti substrate.

2 Experimental Details

The neutron diffraction experiment was carried out by a high-resolution 3-axis setting as schematically shown in Fig. 1. Following the sketch (for small widths of the samples), a maximum resolution can be achieved for minimal dispersion of the whole system. When treated in the momentum space, the orientation of the Δk momentum domains related to the monochromator and analyzer should be matched to that of the sample, and this can be achieved by proper radii of curvature of the bent perfect crystal monochromator and analyzer [3].

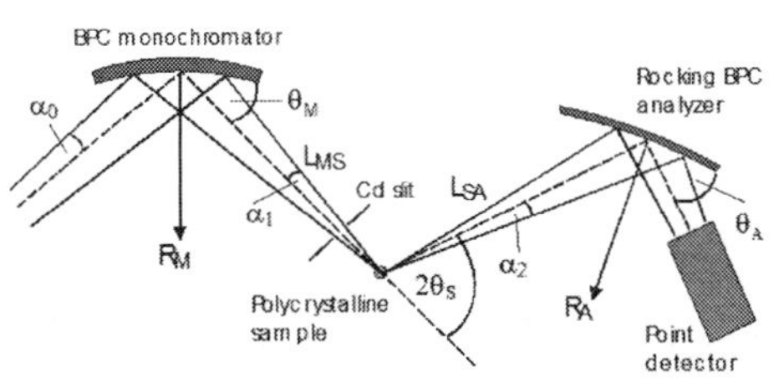

Fig. 1: The scheme of the three-axis diffractometer setting employing BPC monochromator and analyzer as used in the experiment (R_M, R_A – radii of curvature, θ_M, θ_A – Bragg angles.

3 Experimental Results

For the high-resolution measurement we used three samples of TiAl of the cylindrical form of the diameter of $\phi = 9$ mm. Manufacture of specimens was carried out by using a printing device InssTek MX-LAB producing the spot size, the height of the layer and the hatch distance. Laser power, speed of the printing and the dwell time were 130 W, 849 mm/min and 3.5 sec, respectively. Parameters related to the samples were: Both S1 (a preheat of the platform on 500 °C) and S2 (without a preheat) – the 800 mm spot size, the 150 mm height of the layer using the cladding head SDM 800 and the 500 mm hatch distance and S3 (no preheat) – the 400 mm spot size, the 250 mm height of the layer using the cladding head SDM 400 and the 300 mm hatch distance, respectively. The following analyzer rocking curves demonstrate how the differences in preparation of the samples reflect their *FWHM*. It can be stated that the presented unconventional high-resolution neutron diffraction setting can offer additional support to complement the structural information achieved by using the other conventional characterization methodologies and can be interesting for potential external users.

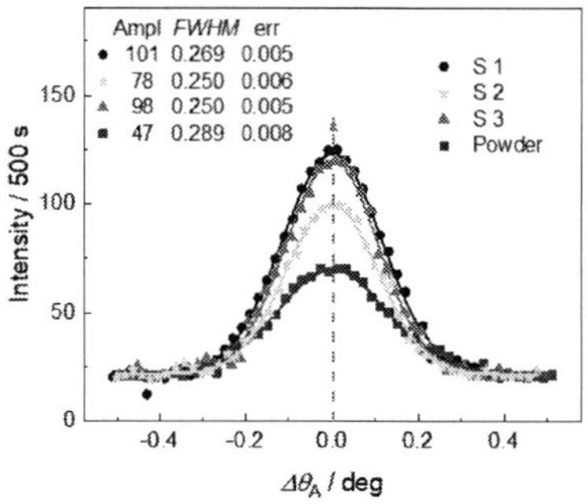

Fig. 2: Analyzer rocking curves related to the individual TiAl samples of the diameter of $\phi = 9$ mm: (a) – sample S1, (b) – sample S2, (c) – sample S3 and (d) – powder sample. The influence of the manufacturing procedure on the *FWHM* of the curves is evident.

Acknowledgements

Measurements were carried out at the CANAM instrument of NPI CAS Řež installed at the CICRR infrastructure, which is financially supported by the Ministry of Education and Culture – project LM2023041. The authors acknowledge support from ESS participation of the Czech Republic – OP (CZ.02.1.01/0.0/0.0/16_013/0001794). Furthermore, they acknowledge support from the CAS in the frame of the program “Strategie AV21, No. 23”. The authors thank B. Michalcová from NPI CAS for significant help with measurements and data elaborations.

References

[1] S.C. Huang, J.C. Chesnutt, Gamma TiAl and its alloys, Intermet Compd. 2 (1994) 73–92.

[2] P. Mikula, V. Ryukhtin, On a possibility of high-resolution neutron diffraction observation of Inconel 718 substructure phases, in proc.: 61st conference on experimental stress analysis EAN (2023), book of extended abstracts, eds. M. Hagara, E. Huňady, Technical University of Košice, Košice, Slovakia.

[3] P. Mikula et al., An alternative neutron diffractometer performance for strain/stress measurements, Powder Diffraction 35 (2020) 185–189, DOI:10.1017/S0885715620000329.

Knowledge Base for Qualitative Assessment of Fatigue Life of Bus Bodywork Joints

R. Minich [1], M. Kepka jr. [1], M. Krizek [1], J. Vaclavik [2], M. Kepka [1,a]

[1] *University of West Bohemia, Univerzitni 8/2732, 301 00 Pilsen, Czech Republic*
[2] *Research and Testing Institute Plzen, Tylova 1581/46, 301 00 Pilsen, Czech Republic*
[a] *kepkam@fst.zcu.cz*

Abstract: The principles of the knowledge base are presented, in which the results of laboratory fatigue tests of welded joints of thin-walled car body profiles are systematically stored. The results from measuring the operating stress of these joints are also included. The collected data can be used in the qualitative assessment of the fatigue life of the bodies of newly developed vehicles (buses, trolleybuses, battery electric buses).

Keywords: knowledge base; bodywork joint; S-N curve; stress spectrum; fatigue life.

1 Introduction

The input data for fatigue life calculations obtained mostly by experiments (laboratory fatigue tests and strain gauge measurements) are:

- S-N curves of assessed construction nodes;
- stress spectra in their critical cross-sections.

Then, fatigue life calculation can be realized using a suitable fatigue damage accumulation rule. This approach can be signed as a quantitative assessment.

The fatigue life can be assessed qualitatively by simply comparing the S-N curve and the aggressiveness of the stress spectrum with historical data stored in the database.

2 Using the Knowledge Base

In the past, the methodology of design of new bus bodyworks, including the fatigue life evaluation, was presented several times in professional literature, recently [1]. The goal of current research and development, carried out as part of doctoral studies, is to create a knowledge base for expert evaluation of the fatigue life of the bodyworks of buses, trolleybuses, and battery electric buses.

The knowledge base is based on a number of laboratory fatigue tests of welded joints of thin-walled bodywork profiles that have already been performed and the results of measuring their operational stresses. The knowledge base contains not only summary evaluations of the performed experiments, i.e., primarily S-N curves and stress spectra, but all source data, including the results of detailed experimental stress analyses and supporting calculations using the finite element method.

The stress spectra are described in the database by the equation proposed in the literature [2]:

$$\log H_i = \left[1 - \left(S_a/S_{a,max}\right)^{\nu}\right] \log H_0 \qquad (1)$$

With decreasing values of ν, the shapes of the amplitude spectra become increasingly hollow (stress spectra are less aggressive in terms of fatigue damage).

For the description of the S-N curves, a relationship using a linear mathematical model, which includes the standard deviation, is suitable [3]:

$$\log(N_f) = logC_0 - s \cdot d - w \cdot \log(S_a) \qquad (2)$$

The standard deviation s of log N_f can be calculated on the basis of experimental data; the value of the standard deviation s can also be found in the mentioned standard [3]. The set of already existing S-N curves is permanently supplemented with results from new projects with industrial partners.

Indicative information on fatigue strength and operational load enables quantitative assessment of structural nodes during the development of new bodies. By comparing it with the information stored in the knowledge base. Fig. 1 shows two boundary situations.

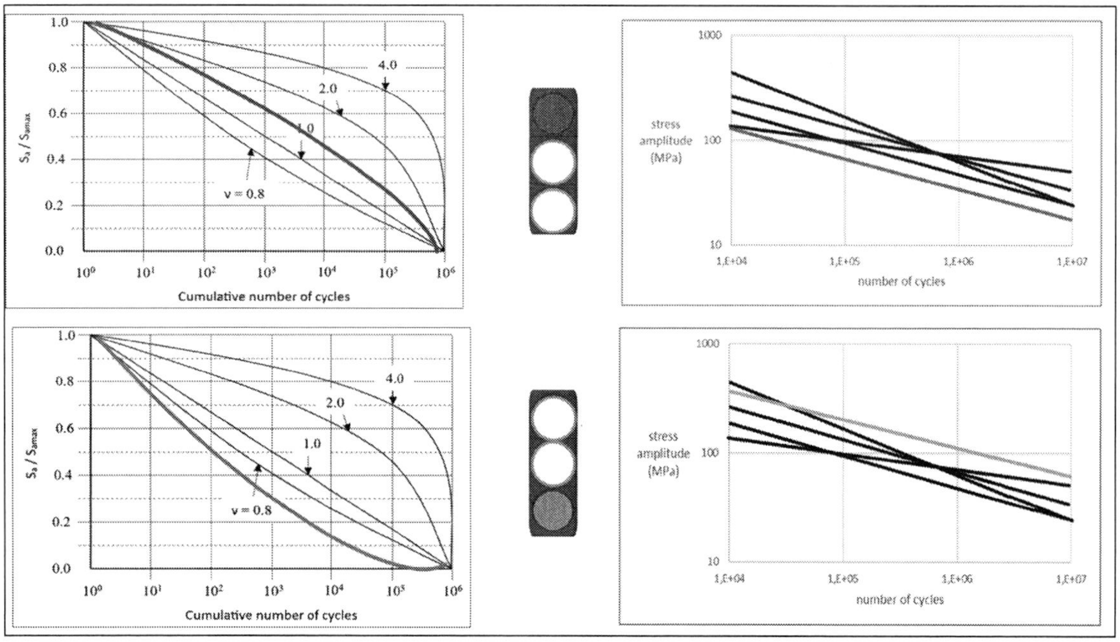

Fig. 1: Quantitative assessment of service fatigue life.

An example is given in green, where the assessed body node is among the high-quality in terms of fatigue life, as evidenced by its position among the S-N curves collected in the database. At the same time, if the load spectrum is not aggressive, the designer of the new bodywork can continue development without fear.

The opposite situation is shown in red. The solution to the new body assembly is obviously not very high quality compared to the previous ones. The load spectrum is also aggressive. In such a case, it is necessary to proceed with structural changes of the body nodes and to think about how to reduce the load level of the vehicle being developed.

3 Conclusion

The advantage of the quantitative approach is that it can rely on estimates of the future load of structural nodes as well as on accelerated procedures for determining their fatigue properties. Nevertheless, such information can be of fundamental importance to the manufacturer of a new vehicle, as it can preliminarily assess the risks of unwanted fatigue damage to critical structural nodes of the newly developed bodywork. And decide, if necessary, on the scope of the detailed assessment and, therefore, its costs.

References

[1] M. Kepka et al., A case study on fatigue life assessment of an electric bus bodywork, International Journal of Fatigue 179 (2024), https://doi.org/10.1016/j.ijfatigue.2023.108040.

[2] P. Heuler, H. Klätschke, Generation and use of standardised load spectra and load–time histories, International Journal of Fatigue 27 (2005), https://doi.org/10.1016/j.ijfatigue.2004.09.012.

[3] BS 7608:1993 – Fatigue design and assessment of steel structures.

Strains in Electronic Assembly of Printed Circuit Boards

R. Musilová [1,a], J. Václavík [1]

[1] *Research and Testing Institute Plzen, Tylova 1581/46, 301 00 Plzeň, Czech Republic*
[a] *vaclavik@vzuplzen.cz*

Abstract: This article describes the methodology of testing PCBA during assembly process.

Keywords: PCBA manufacturing process; strain gauge test on PCBA; strain rate.

1 Introduction

Strain gage testing quantifiably analyses the strain and strain rate levels that the electronic components are subjected to during Printed Circuit Board Assembly (SMT & THT), final product assembly and test. It represents complex work including FEA analysis, strain gauges installation and measurement and sometimes metallographic analysis of the cracked soldered joint, which often is the initiation for the own strain investigation. In addition, investigations of strain levels in the laboratory are sometimes necessary. This testing can improve manufacturing operations, that can pose high risk for interconnect damage. Excessive strains can result in solder ball cracking, trace damage, pad lifting and package substrate cracking.

The guidelines for the tests are described in JEDEC methodology [1]. Although this standard is focused mainly on surface mount devices, the test approach can be used also for non-area-discrete SMT devices. The strain limits, however, are specified in non-obligatory IPC-WP-01 white paper only and have to be stated always by the customer. Board flexure control using strain gage measurement is the accepted method to verify that the assembly and test processes are not exceeding defined strain levels [2].

Typically, two test boards are instrumented, one after reflow with SMT components only, the other after wave solder with both SMT and trough-hole components. Appropriate wire management, bundling and securing the wires is important. The wires should be run between components, where they will not interfere with any process steps.

The main objective of PCB tests is to determine the principal strains in the areas of high strain concentration especially at the corners of large circuit components [3]. The strain gauge rosettes with grid length smaller than 1 mm are used for this purpose.

2 Results

Different time strain histories are measured during different manufacturing steps - solder printing, components deposition and PCB cutting are shown in Fig. 1 from top to down.

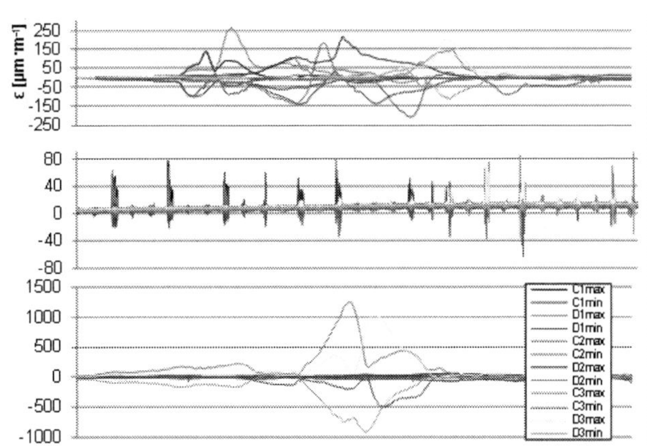

Fig. 1: Resulting strains, measured during some PCB operation steps.

Another task is strain optimization during the assembly of the finished PCB to its case. Details of strain – time history during assembly of the PCB to the case of some control unit is given in Fig. 2. Here, the evaluated maximum principal strain and sheer strain and their rates are given.

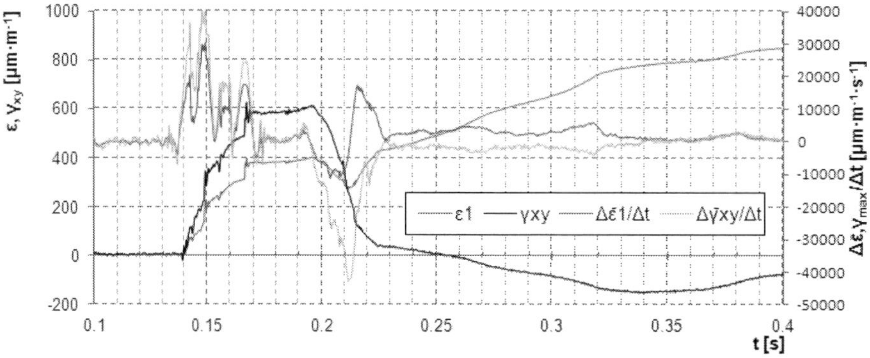

Fig. 2: Maximum principal strain ε1, sheer strain γ_{xy} and their strain rates.

The allowable strain is the function of PCB thickness and the stress rate. In Fig. 3, the distribution of strains is related to allowable value inclusive of the strain rate (the allowable strain lowers with increasing rate).

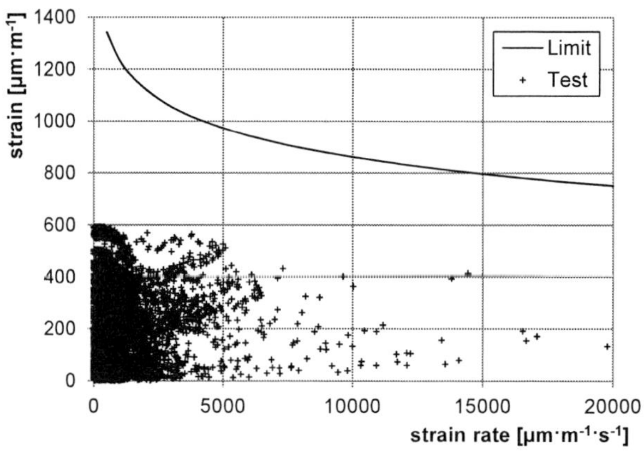

Fig. 3: Measured strain vs. allowable depending on strain rate.

3 Conclusion

The purpose of the article is to show the reason and importance of PCB tests. More analysis and examples including the stain level investigations in the laboratory are presented in the full text of the article.

Acknowledgement

The article originated in the framework of the institutional support for the long-term conceptual development of the research organization.

References

[1] Printed Circuit Assembly Strain Gauge Test Guideline, IPC/JEDEC–9704A (February 2012).

[2] J. Chvojan, J. Václavík, PCB Tests during Assembly and Splitting, in proc.: 8th International Conference on Experimental Mechanics (ICEM18) (2018), Brussels, Belgium.

[3] M. Lyczek, W. Skarka, Review of Methods for PCB Panel Depanelization and Methods for Correct Assembly of Electronic Components on PCB Panels, Electronics 13 (2024) 1255, https://doi.org/10.3390/electronics13071255.

Shrinkage of Rammed Unfired Earth in Long Term Observation

P. Padevět [1,a], B. Mužíková [1]

[1] *Czech Technical University in Prague, Faculty of Civil Engineering, Department of Mechanics, Thákurova 7, Prague 6, 166 29, Czech Republic*

[a] *pavel.padevet@fsv.cvut.cz*

Abstract: The paper is devoted to the creep and shrinkage of compacted unfired earth. The creep of illitic-kaolinitic clay has been monitored for a long time. The initial measurement was focused on the initial course of creep. The long-term measurement took place for almost 4 years. The compacted clay was prepared using a water coefficient of 0.295, when the amount of clay used in the mixture was 30%.

Keywords: rammed unfired earth; illitic kaolinitic clay; shrinkage; creeping of material; water ratio.

1 Introduction

For construction purposes, rammed unfired earth is a material for which it is advisable to focus not only on strength properties, but also on rheological characteristics. An important factor for monitoring the rheology of unfired clay is the need for mixing water during the creation of the mass. Creep and shrinkage are rheological properties of clay that have a significant effect on the long-term behavior of the structure.

2 Tested Material

The pressed clay was prepared from sand, illitic kaolinitic clay and water. The mass ratio of water to illitic kaolinitic clay was 0.295. The ratio of sand and clay was 70:30. The samples for creep and shrinkage testing were prepared together with the samples intended for material tests.

Fig. 1: One of the specimens of illitic kaolinitic clay after creep test.

The creep specimens had a hexagon-shaped base with a distance of 20 mm from opposite edges. The height of the specimens was 70 mm due to the installation of the lever measuring mechanism, as is possible seen in Fig. 1. The measurement was started on 6 specimens, where 3 samples were intended for creep measurement and 3 samples for shrinkage measurement.

3 Instrumentation of Experiment

The creep specimens had a hexagon-shaped base with a distance of 20 mm from opposite edges. The height of the specimens was 70 mm. All tests took place in lever mechanisms. A lever mechanism for measuring creep applied a load to the specimen of 53 N. Shrinkage was measured in a lever device that was used only to clamp the strain sensors. These bodies were not subjected to force, as in the case of creep measurements. The size of creep and shrinkage was recorded using optoelectronic probes SM 12. 3 sensors were placed around each body in the axial direction of loading of the samples.

The specimens were immediately inserted into the test devices and started with the measurement. The data has been collected in a locally placed PC file. During the measurement, the data from one body with the measurement of complete and one body with a shrink measurement was terminated.

4 Conclusion

Long-term measurement managed to obtain a unique result of the completion and shrinkage of notched unbalanced clay with a water coefficient of 0.295. Similar results, but in the shorter period of measurement, you can see in Fig. 2.

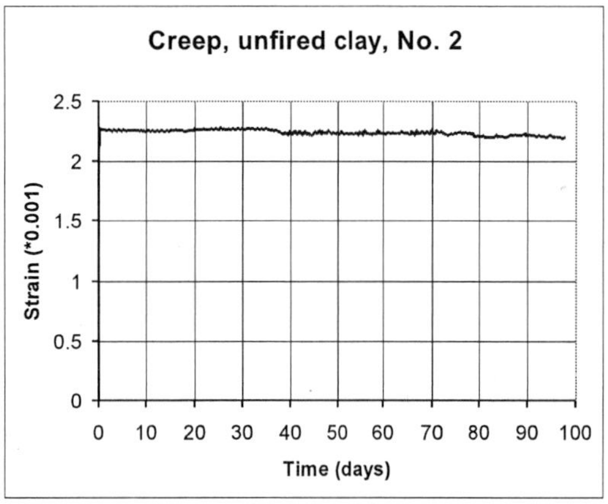

a) Creep curve of rammed earth prepared form illitic kaolinitic clay, water ratio 0.37

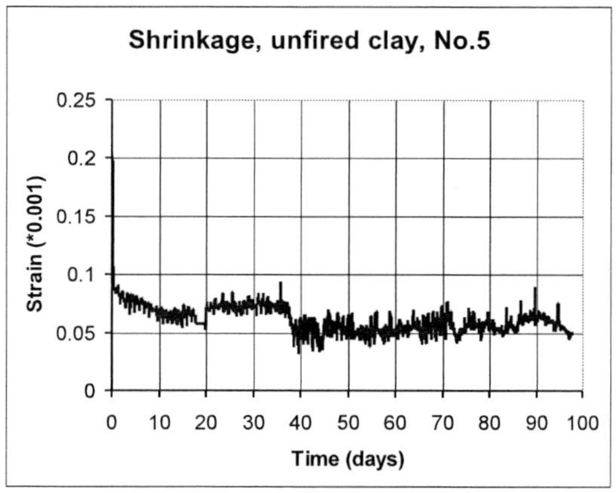

b) Shrinkage curve of rammed earth prepared form illitic kaolinitic clay, water ratio 0.37

Fig. 2: Curves of creep and shrinkage for rammed unfired earth with ratio of clay and sand 20:80 and water ratio 0.37.

Acknowledgement

The paper was supported of the project CTU No. SGS 24/037/OHK1/1T/11.

References

[1] P. Padevět, P. Bittnar, B. Mužíková, Comparison of shrinkage of the rammed earth prepared from illitic and illitic kaolinitic clay, in proc: 57th International Scientific Conference on Experimental Stress Analysis (2019), 356–360, ISBN 978-802145766-9.

Analysis of the Strain-Rate Dependence on the Mechanical Properties of Polypropylene

P. Palička [1,a], R. Huňady [1], M. Hančinová [1], E. Dubňanská [1]

[1] *Technical University of Košice; Letná 1/9, Košice, Slovakia; peter.palicka@tuke.sk, robert.hunady@tuke.sk, martina.hancinova@tuke.sk, erika.dubnanska@tuke.sk*
[a] *peter.palicka@tuke.sk*

Abstract: Since the end of the 20th century, the use of plastics has undoubtedly been expanding, not only in technical applications, but also in the food industry, medicine, and many other sectors. The aim of material development is to achieve the desired properties depending on its specific application. The mechanical behaviour of polymeric materials is significantly influenced by temperature, strain-rate and their processing. In general, polymers behave as viscoelastic materials and exhibit intermediate behaviour between elastic solids and viscous liquids. This paper is focused on investigating the effect of strain-rate on the mechanical properties of polypropylene. Flexural and uniaxial tensile tests at three strain-rates were carried out as part of the material testing.

Keywords: polymer; thermoplastic; polypropylene; viscoelasticity; strain-rate.

1 Introduction

The same parameters are used to characterise the mechanical properties of plastics as for metals, i.e. Young's modulus, Yield stress, Ultimate strength, Poisson's ratio, but their mechanical behaviour differs considerably to metals. In comparison to other materials, the mechanical response of plastics varies to a much greater extent depending on the strain rate, temperature, and the external environment. Polymers at high loading frequency or low temperature behave like glassy substances, while at low loading frequency or high temperature polymers behave like rubbery materials. In the first case, the mechanical properties are higher, but the polymer becomes brittle, while in the second case the mechanical properties are lower and the polymer becomes ductile, i.e. it transitions to a rubber-like state [1].

2 Experimental Investigation

The investigated material was polypropylene filled with 12% talc. Uniaxial tensile and flexural tests were performed on the universal machine LabTest 5.050ST. Dynamic mechanical analysis (DMA) was also performed to find the frequency dependence of the flexural modulus. The DMA analysis was performed in a three-point bend at three frequencies 0.4, 4 and 40 Hz. Tensile tests were performed on 7 specimens at strain-rates of 10-2, 100 and 102 s^{-1}. Three-point flexural and DMA tests were realized on 5 test specimens. The tensile tests were carried out up to failure and the flexural tests were carried out only in the elastic region. Deformation in tensile tests was measured by using a video extensometer from Xsight. This video extensometer measures deformation by using digital correlation technology. At strain-rates of 100 and 102 s^{-1}, it was necessary to correct the tensile curve to a storage modulus that was obtained from the DMA tests due to the small amount of collected data. This correction was carried out in Matlab software, where the measured tensile curves were approximated so that the tensile modulus corresponded to the storage modulus. The tensile test specimen was a conventional "dogbone" specimen with 2.5 mm thickness. The specimen for flexural and DMA tests was in the shape of a rectangle with the same 2.5 mm thickness. The specimens were of small size due to the high loading rates. For large specimens measured at high loading rates, undesirable effects such as specimen flutter occur, which directly affect the measurement results. The tensile and flexural test specimens are shown in Fig. 1.

a) Uniaxial tensile test. b) Three-point flexural test.

Fig. 1: Material tests.

Fig. 2 shows the material test results from measurements with 10^{-2} strain-rate. From the plots, the stress-strain curves correspond well in the elastic region but differ significantly in the plastic region at failure.

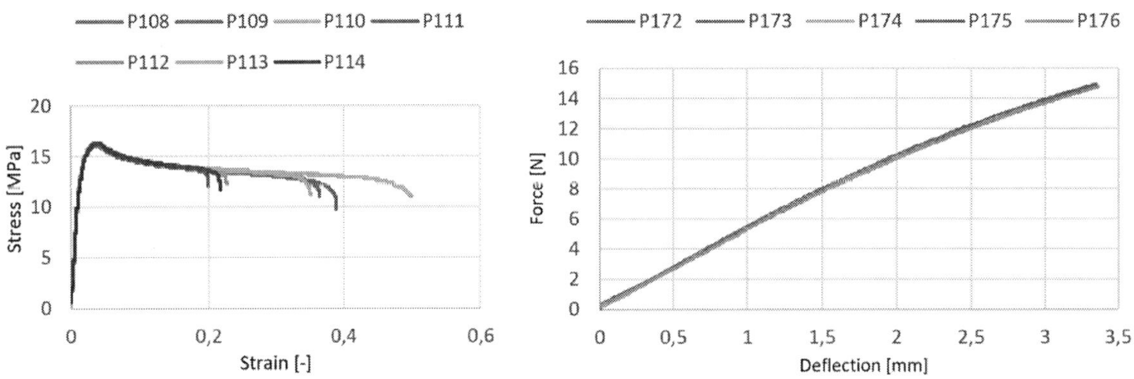

a) Uniaxial tensile test. b) Three-point flexural test.

Fig. 2: Material test results.

Acknowledgement

This work was supported by the Science Grant Agency of the Ministry of Education, Science, Research and Sport of the Slovak Republic and the Slovak Academy of Sciences under the project VEGA 1/0500/20.

References

[1] K. Deshmukh et al., Chapter 4 – Mechanical analysis of polymers, Polymer Science and Innovative Applications, Elsevier, 2020, pp. 117–152, ISBN 9780128168080, https://doi.org/10.1016/B978-0-12-816808-0.00004-4.

Determination of Shear Properties of 3D Printed Composite

P. Pavlíček [1,a], M. Fusek [1], M. Šofer [1]

[1] *Technical University of Ostrava, Czech Republic*
[a] *pav0151@vsb.cz*

Abstract: Determination of mechanical properties in shear is important for defining of material model of composite. In this work are compared two types of shear experiment methods when every method give different result. Understanding behaviour of composite material under shear loading gave information about the usability of the tested method and for future research.

Keywords: 3D printing; composites; material model; DIC.

1 Introduction

This work is focused on the methodology of testing 3D printed composites and obtain mechanical properties of composite under shear loading. Behaviour under shear loading is important for define material model of composite. In this work is main goal compare two experimental methods:

- ISO 527-4 [1],
- ASTM 5379 [2].

These two standards define test parameter for shear testing when both methods gave different results. Understanding of behaviour composite under shear loading in different tests conditions give information for the next research of test methods.

3D printed composite in comparison with traditional composites has more inhomogeneous structure. This type of structure in some cases of loading causes dispersion of the data.

2 Test Procedure

Composites were tested on a universal tensile machine Testometric MC-500 with 50 kN force sensor. The test procedure was done in with accordance ISO527-4 and ASTM 5379. Samples before testing were modified when must necessarily remove the first layer of onyx. This layer from the research has a significant influence on load response and causes distortion of results. The deformation of the sample was measured with the DIC method. In Figs. 1 and 2 below are presented the tested methods.

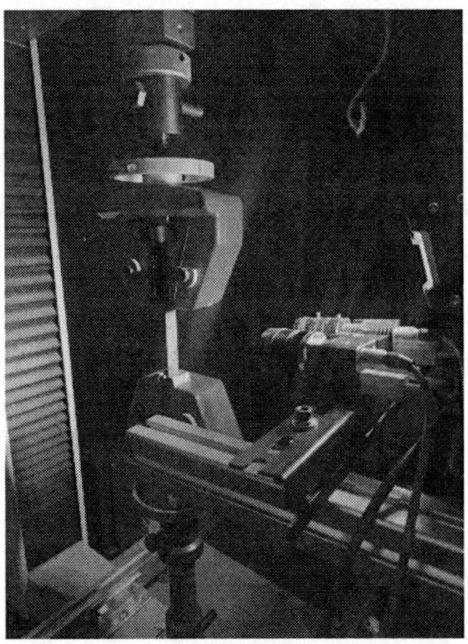

Fig. 1: ISO 527-4.

Fig. 2: ASTM 5379.

Results

From the experiment was obtained test data which are written in Tab. 1 and graphs on Figs. 3 and 4. Data was evaluated in with accordance ISO 2602 [3].

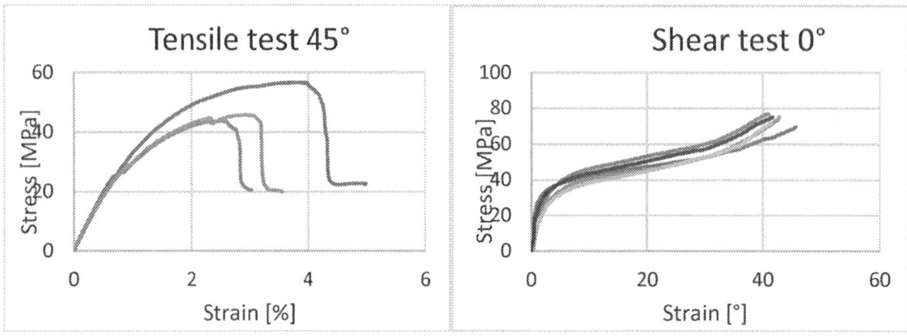

Fig. 3 and Fig. 4

Tab. 1

Property	ISO 527-4	ASTM 5379	Difference [%]	Unit
Young´s modulus	3562 ± 142	2671 ± 172	35	MPa
Strength	48 ± 5.6	62 ± 4.4	29	MPa
Elongation	3.86 ± 0.8	36 ± 3.8	947	%
Yield Stress	26.6 ± 2.9	15.5 ± 2.2	42	MPa

From the results is a visible difference in the graphs when ASTM D3039 gave higher results than the ISO 527-4. These results are mainly visible in elongation when ASTM D3039 has a different type of failure than ISO 527-4 test.

3 Conclusion

From the results presented above was not found a correlation between ASTM D3039 and ISO 527.4 both methods gave a different results. The mean error between these methods is around 35%. The main difference between these two types of tests is in the different types of failure and stress distribution.

Acknowledgement

This work was supported by Specific Research "Experimental and Numerical Modeling for Solving Problems in Mechanics and Biomechanics" (SP2024/037).

References

[1] ISO 527-4, Plasty – Stanovení tahových vlastností – Část 4: Zkušební podmínky pro izotropní a ortotropní plastové kompozity vyztužené vlákny, Praha, Úřad pro technickou normalizaci, metrologii a státní zkušebnictví, 2023, 44 s., třídící znak 640604.

[2] ASTM D3039, Standard Test Method for Tensile Properties of Polymer Matrix Composite Materials, 13 p., USA, DOI:10.1520/D3039_D3039M-08.

[3] ISO 2602, Statistická interpretace výsledků zkoušek. Odhad průměru. Konfidenční interval, Praha, Úřad pro technickou normalizaci, metrologii a státní zkušebnictví, 1993, 12 s., třídící znak 010231.

Kyowa MRS-100 Digital Telemetry System

"Make your measurement wireless"

K. Pogány [1,a], A. Sarusi-Kiss [1]

[1] *KESACON Kft., Hungary*
[a] *pogany@kesacon.com*

Abstract: Our paper demonstrates how our company – KESACON – has encountered a strain measurement technical issue at an important partner in the automotive electronics industry. Strain measurement is basically implemented by using strain gages with attached cables, where cables – normally several meters long - are connected to the measurement device. Nevertheless, life is not always so simple. This given, somewhat special measurement environment – a moving robot arm – simply made it impossible to handle the cable of the strain gage as the cable itself would have been destroyed by the moving robot arm. So KESACON has exploited Kyowa's newest technology: we have taken our Kyowa MRS-100 telemetry system with our Kyowa strain gages and have easily solved this measurement task.

Keywords: wireless; strain gage; robot arm; telemetry; strain measurement.

1 Introduction

The telemetry system can measure rotating, moving objects by establishing a wireless connection between the measuring instrument and the sensor. Kyowa MRS-100 series includes a small-sized transmitter unit for narrow places, which communicates stability by using digital modulation. The communication range is max. 30 meters (line of sight distance) but depends on the measurement environment. Applicable sensors: strain-gage, strain gage based sensors & voltage-output sensors. The system supports a quarter bridge & full bridge system (half-bridge available as customization). After communication is established by the telemetry system, a datalogger is always needed for recording the data.

2 Implemented Measurement Task

2.1 Environment

Our Czech partner – the automotive electronics manufacturing industry – for whom we are supplying Kyowa strain gages, had to measure on printed board during the assembly phase and production. That seemed to be simple. Nevertheless, they had to measure and analyze the stress impact of a moving robot arm on the PCB. Implementation was impossible with cabled strain gages. We suggested them wireless way of strain measurement with a telemetry system, eliminating strain gage cables. They had their own measurement device but their system was not able to handle this task. KESACON introduced them Kyowa MRS-100 telemetry system.

2.2 Technical Set-up

For establishing the wireless communication, KESACON has used own MRS-100 telemetry system, which includes 1 x 4-ch *strain transmitter unit* with a quarter bridge adapter and 1 x 4-ch *receiver unit*. The small-sized transmitter we fixed onto the robot arm, which was holding the PCB (see Fig. 1). The board contained 1 x rosette and 1 x uni-axial Kyowa strain gage (4 channels), all with short, 10 – 15 cm long narrow cable, which was led along the robot arm to reach the bridge adapter of the transmitter unit. The receiver unit and datalogger (Kyowa EDX-200A) were 5 meters behind a protective glass wall and were monitoring the entire assembly process by logging the data (see Fig. 2.).

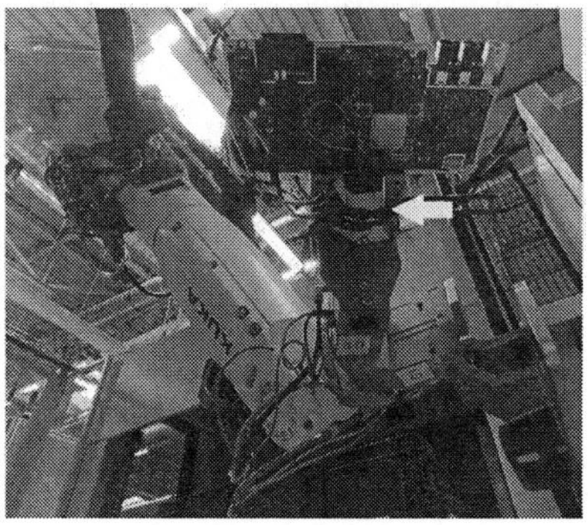

Fig. 1: Robot arm working with the PCB.

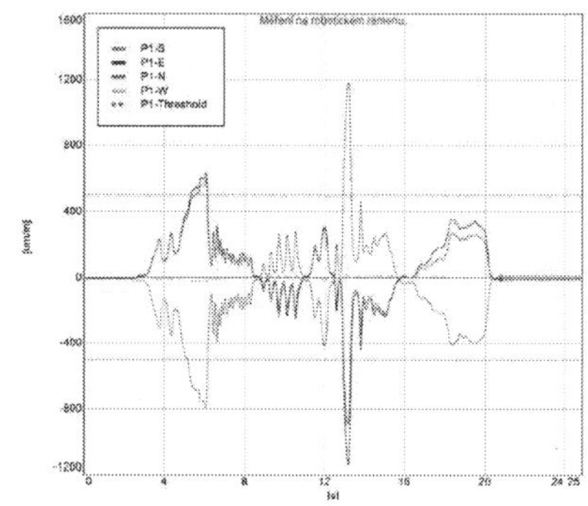

Fig. 2: Graphic evaluation of the PCB.

2.3 Result

The measurement has been successfully implemented with the involvement of our telemetry system and by this, the data recording part has been implemented in the usual way. Our partner was satisfied with the technology, process and results although the project first seemed hard first from implementability aspects. Later on they ordered the telemetry system.

2.4 Further Application References

As a further reference, KESACON has measured the torque on a rotating shaft of an agricultural machine in Poland. The transmitter has a centrifugal acceleration resistance of up to 3000 G (29420 m/s^2). Another example is a completed measurement on the shaft of a railway carriage in Romania at a national railway authority. They already have 2 units of this telemetry system.

3 Conclusion

Kyowa MRS-100 telemetry system is your ideal solution for those strain measurement projects where you meet strain gage cabling issues due to narrow places or moving/rotating objects. The system solves the measurement of high-speed phenomena in experimental, research, manufacturing, automotive fields and transmits the test data in almost real-time. You can use the system not only for strain gages but for load/pressure/acceleration/torque sensors and thermocouples wirelessly for quality controlling, improving your productivity in a wireless way. See some application ideas below. Just try it, take and exploit it!!!

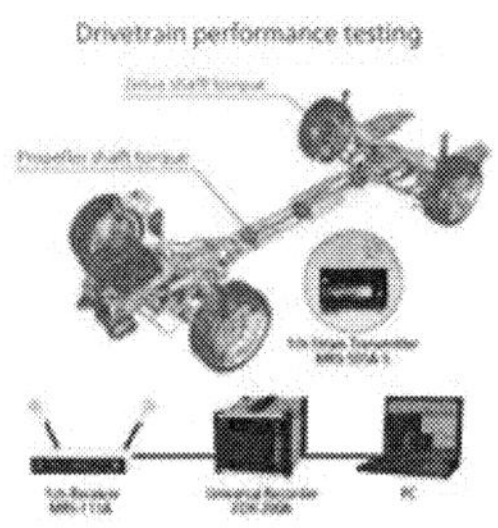

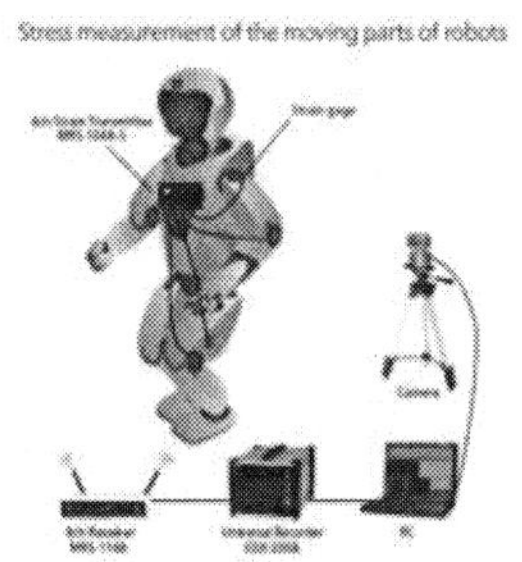

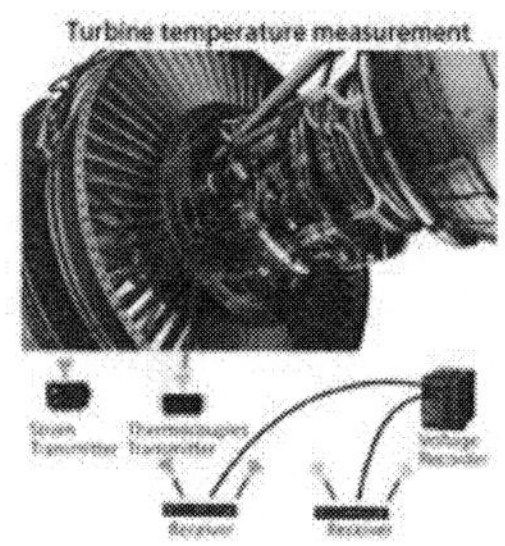

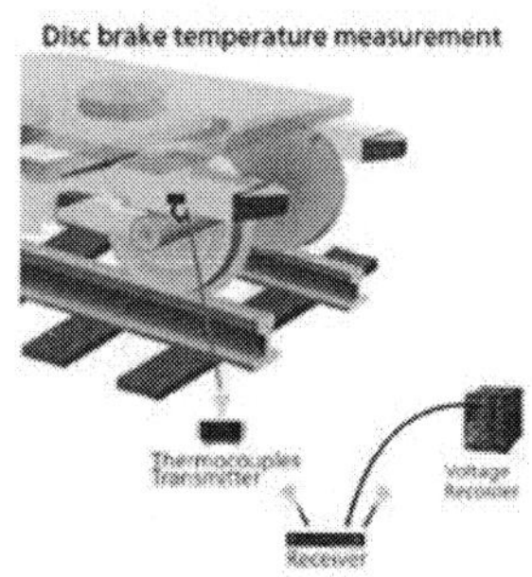

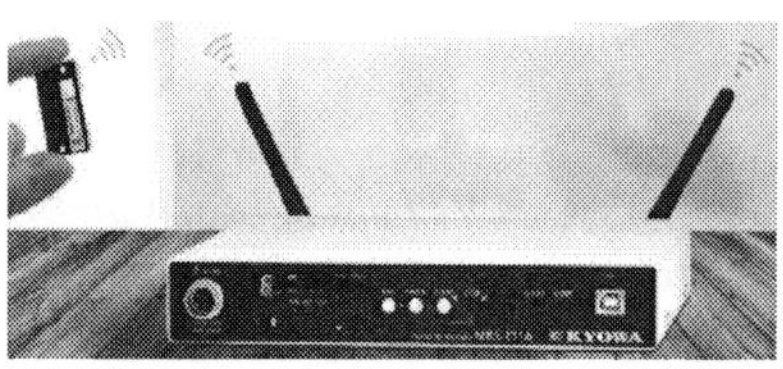

References

[1] KESACON Kft., Own wireless strain measurement projects (ing. Kristián Pogány, Sarusi-Kiss Andrea).

[2] Kyowa Electronic Instruments Co., Ltd. – MRS-100 series digital telemeter catalog.

[3] Kyowa Electronic Instruments Co., Ltd. – https://product.kyowa-ei.com/en/special/topics/mrs-100a.

Characterization of the Interfacial Transition Zone between the Cement Matrix and Glass Powder Particles

Z. Prošek [1,2,a], A. Palička[1], P. Tesárek [1]

[1] *Department of Mechanics, Faculty of Civil Engineering, CTU – Czech Technical University in Prague, Thákurova 7, 166 29 Prague, Czech Republic*

[2] *University Centre for Energy Efficient Buildings, CTU – Czech Technical University in Prague, Třinecká 1024, 273 43 Buštěhrad, Czech Republic*

[a] *zdenek.prosek@fsv.cvut.cz*

Abstract: The article deals with the determination of the width of the interfacial transition zone (ITZ) between the waste glass powder (WGP) and the cement matrix. The transition zone negatively affects the resulting mechanical properties of the composite. The article deals with surface treatments using plasma treatments under different gases to reduce the size of the ITZ, namely oxygen (O_2) and hydrogen (H_2) plasma. ITZ is measured using a scanning electron microscope (SEM) in combination with the determination of the percentage representation of individual elements by using energy dispersive spectroscopy (EDS). The results are compared with mechanical tests on samples of 20 × 20 × 100 mm size. As part of the article, a reference sample without surface treatment was also measured. The results show that using oxygen plasma it is possible to reduce the ITZ and thereby also improve the mechanical properties of the resulting cement composites.

Keywords: interfacial transition zone; waste glass particles; cement matrix; scanning electron microscope; mechanical properties.

1 Introduction

Cement composites are made up of three main components: matrix, aggregate, and interfacial transition zones (ITZ) [1]. Furthermore, the cement matrix consists of C-S-H gel, unhydrated clinkers, and portlandite (CH). The ITZ is acknowledged as the most vulnerable segment of cement composites, resulting in a reduced compressive strength compared to aggregate and hardened cement paste [2]. Furthermore, the presence of micro cracks within the ITZ contributes to changing the response of cement-based materials from linear to non-linear under loading conditions [3]. Mechanically, the extent of micro cracks within the ITZ is dictated by the degree of strain mismatch between the aggregate and the paste [4]. When stress is evenly distributed, differences in strain arise due to variations in the elastic modulus, the shear modulus of the aggregate, the matrix, ITZ, and the stress concentration. Serving as a bridge between the matrix and the aggregate, if the ITZ does not transmit stress effectively between phases, the compressive strength of the cement composite is compromised, even if one phase exhibits greater stiffness [5].

To understand the mechanism of formation of ITZ and its impact on the properties of cement composites, a detailed microstructural analysis was performed. It has been observed that the presence of CH, along with relatively high porosity, respectively, low-density C-S-H gel (LD CSH) and large capillary pores, significantly influences mechanical properties. Furthermore, the presence of large CH particles weakens van der Waals forces and has adverse effects on mechanical properties [6]. Apart from CH, micro cracks and high porosity are prone to forming within the ITZ, further diminishing mechanical properties. Another indicative factor for ITZ formation is the calcium-to-silicon (Ca/Si) ratio in the cement matrix. A ratio of 0.5-1.5 results in the formation of high-density C-S-H gel (HD CSH), while a ratio of 1.5-2 yields LD CSH [7]. Another significant effect is the high content of amorphous SiO_2 in WGP, exceeding 70%, which, when exposed to an alkaline environment, can dissolve and generate a secondary C-S-H gel [8]. This pozzolanic reaction occurs gradually, enhancing the strength of the samples at later stages [9].

The focus of this article lies on the utilisation of WGP within cement paste, with the aim of enhancing its activation through surface plasma treatments. The objective is to decrease the thickness of the ITZ between WGP and the cement matrix, thereby leveraging the potential for the secondary formation of C-S-H gels. This approach aims to create a compact structure with a higher strength value compared to the reference sample.

2 Results

The samples were composed of 80 wt. % Portland cement CEM I 42.5R (Radotín) and 20 wt. % of the glass powder waste, which comes from the waste material of the Recifa Packaging Industry, as. The powder was surface treated with plasma of O_2 and H2. The O_2 plasma made the surface of the WGP hydrophilic and thus reduced the ITZ, and the 28-day samples achieved 13% higher flexural strength than the reference sample. After 126 days, the differences were even greater and the compressive strength of the sample where the WGP surface was treated with plasma O2 reached values 6% higher than the reference.

Acknowledgement

This paper was financially supported by the GA CR research project under the number GA23-05500S and by Czech Technical University in Prague under No. SGS project SGS24/037/OHK1/1T/11. The authors also thank Recifa a.s.

References

[1] A. M. Neville, Properties of Concrete, London, Longman, 1995.

[2] Y. Ke et al., Identification of microstructural characteristics in lightweight aggregate concretes by micromechanical modelling including the interfacial transition zone (ITZ). Cement and Concrete Research 40 (2010) 1590–1600, https://doi.org/10.1016/j.cemconres.2010.07.001.

[3] T. Akçaoğlu, T. Mustafa, T. Çelik, Effect of coarse aggregate size and matrix quality on ITZ and failure behavior of concrete under uniaxial compression, Cement and Concrete Composites 26 (2004) 633–638, https://doi.org/10.1016/S0958-9465(03)00092-1.

[4] G. Prokopski, B. Langier, Effect of water/cement ratio and silica fume addition on the fracture toughness and morphology of fractured surfaces of gravel concretes, Cement and Concrete Research 30 (2000) 1427–1433, https://doi.org/10.1016/S0008-8846(00)00332-X.

[5] J. A. Larbi, The cement paste-aggregate interfacial zone in concrete, Doctoral thesis, Delft, TU Delft, 1991.

[6] N. Van Tuan et al., Hydration and microstructure of ultra high performance concrete incorporating rice husk ash, Cement and Concrete Research 41 (2011) 1104–1111, https://doi.org/10.1016/j.cemconres.2011.06.009.

[7] Z. Prošek, V. Nežerka, P. Tesárek, Enhancing cementitious pastes with waste marble sludge, Construction and Building Materials 255 (2020) 119372, https://doi.org/10.1016/j.conbuildmat.2020.119372.

[8] Y. Kong et al., SEM analysis of the interfacial transition zone between cement-glass powder paste and aggregate of mortar under microwave curing, Materials 9 (2016) 733, https://doi.org/10.3390/ma9090733.

[9] H. Du, K. H. Tan, Concrete with recycled glass as fine aggregates. ACI Materials Journal 111 (2014) 47–58, http://dx.doi.org/10.14359/51686446.

Bending Analysis of GFRP Composite Plate with a Hole Using Carbon Fiber Sensors

N. Schmidová [1,a], K. Doubrava [1], Z. Padovec [1], D. Blaha [1], R. Kottner [1,2], C. Novotný[1], M. Růžička [1]

[1] *Division of Strength and Elasticity, Faculty of Mechanical Engineering, Czech Technical University in Prague, Technická 4, 166 07 Prague, Czech Republic*
[2] *ACO Industries Tabor s.r.o., Průmyslová 1158, 391 01 Sezimovo Ústí, Česká republika*
[a] *Nikola.Schmidova@cvut.cz*

Abstract: Glass fiber composite vessels are used for underground water storage. During transport and operation of such vessels damage occurs. This paper investigates the possibility of utilization of integrated Carbon Fiber Sensors (CFSs) for Structural Health Monitoring (SHM) of such glass fiber composite vessels. In this study, we performed a flexural analysis of a glass fibre-reinforced composite panel with a central hole. The strain monitoring using CFSs near the hole was performed. The obtained results confirm the possibility of using CFSs for SHM of GFRP vessels.

Keywords: Structural Health Monitoring; carbon fiber; sensor; Glass Fiber Composite vessel.

1 Introduction

A system of drainage channels and retention tanks is usually used for the collection and eventual subsequent treatment of surface stormwater from large paved areas, allowing the water management cycle to be implemented from water capture, through treatment to final retention or recycling.

The cylindrical glass fiber composite wound vessels of interest are used to collect these waters. Vessels of various volumes are placed usually underground vertically or horizontally. Increased loads and stresses on the vessel walls, particularly localized loads, may occur during the placement of the vessels in the backfill and during actual operation, which may cause damage to the integrity of the vessel. It would therefore be desirable to develop a system for the possible indication of such local stresses, or of disruption of the vessel composite structure itself.

One of the critical areas of the vessel is the area around the mouth of the GFRP vessel. Sensors made of carbon fiber tow (Carbon Fiber Sensors = CFSs) integrated during the manufacturing process can be used for structural monitoring of the vessel.

An initial measurement using CFSs was performed during a flexural test of the plate-type specimen with a central hole.

2 Experimental Procedure

The flexural test of the GFRP plate with a central hole was performed in the configuration depicted in [1]. Three specimens were investigated. The specimens were manufactured using an automated filament winding procedure of glass fibre tows with polyester resin. The length of the specimen was 400 mm, the width was 200 mm and the thickness was 9.76 mm. The central hole had a diameter of 80 mm. The specimens were cut out from a winded vessel.

The CFSs were manufactured according to the procedure published in [1] in a length of 80 mm. The CFSs were attached to the surface of the specimen after manufacturing and covered with a covering layer. The three CFSs were placed along the longitudinal axes of the specimen according to the Fig. 1.

Fig. 1: Configuration of the flexural test of the specimen with a central hole and integrated Carbon Fiber Sensors.

3 Results and Conclusion

The results obtained from the strain measurements using CFSs show good repeatability and are in good agreement with the expected strain distribution based on the theory and FEM analysis.

Acknowledgement

The project TN02000010 of National Competence Centre of Mechatronics and Smart Technologies for Mechanical Engineering is co-financed from the state budget by the Technology Agency of the Czech Republic within the National Centres of Competence Programme.

References

[1] A. Horoschenkoff, C. Christner, Carbon Fibre Sensor: Theory and Application, in book: Composites and Their Applications (2012), ed. N. Hu, IntechOpen, online, DOI: 10.5772/50504.

Validation of Simulation of the Temperature Field of a Part during Selective Laser Melting Procedure

R. Skácelík [1,a] , M. Španiel [1]

[1] *Department of Mechanics, Biomechanics and Mechatronics, Faculty of Mechanical Engineering, Czech Technical University in Prague, Technická 4, 160 00 Prague 6, Czech Republic*

[a] *radek.skacelik@fs.cvut.cz*

Abstract: The heat that is supplied to the printed part during the selective melting procedure must be dissipated so that the part does not overheat. The procedure fulfilling this requirement can be effectively designed using validated numerical simulations. The measurement of the temperature field on the top layer of the part using a thermocamera placed in the area of printing is the main topic of the contribution. The measured temperature field is compared with one obtained from simulation employing the finite element method. The comparison provides us with the validation of both the finite element model and computational procedure which are described in the article as well.

Keywords: 3d printing; thermal camera; finale element method; recouting; selective laser melting.

1 Introduction

Selective laser melting is one of the additive technologies based on melting metal powder. The principle consists of consecutive application of thin layers of metal powder in workspace that is locally melted at the sites of the printed part. This creates a solid occupying the entire geometry of the printed part while the rest of the workspace remains filled with powder [1]. The processing parameters, i.e. laser input power, scanning speed and slit pitch, have a significant effect on the heat supplied and, consequently on the temperature of the part. This heat is dissipated through the part of the platform. The part is subjected to dimensional changes due to the thermal expansion of the material. This reduces the quality and the consistency of the thickness of the new layers applied. In extreme cases, this can result in the closure of the gap between the part and the coating device, causing print failure. Such accidents reduce the efficiency of the entire process and therefore it is an effort to be able to predict them before the actual printing of the part takes place.

2 Methods

For validation, a 10 mm diameter cylinder with a 45 degree inclination cone was chosen. Such a specimen was placed centrally on the base of the printing surface so that its behaviour was not affected by the positioning. A TRIUMF 1000 printer with a 100 mm diameter circular print base was used for printing. The print area was scanned with a thermal imaging camera at the moment before a new layer of powder was applied (Fig. 1) so that the temperature of the part at the moment of the new layer was determined. During the formation of the new layer, the camera lens was covered by a servo-controlled aperture.

2.1 Experimental Measurement

The measurements were performed with a FLIR lepton 3.5 thermal imaging camera whose compact dimensions allow it to be placed close to the printing area. The pixel density of the measurement is about 1 pixel per square millimeter. The camera was calibrated using the method described in [1] by modifying the heating of the model in the annealing furnace. The measured values were validated against the thermocouple values. The measurement point was chosen in the middle of the part and the temperature values were averaged using the surrounding pixel values. The raw data was transferred from the thermal camera directly to the computer where the data was processed and averaged.

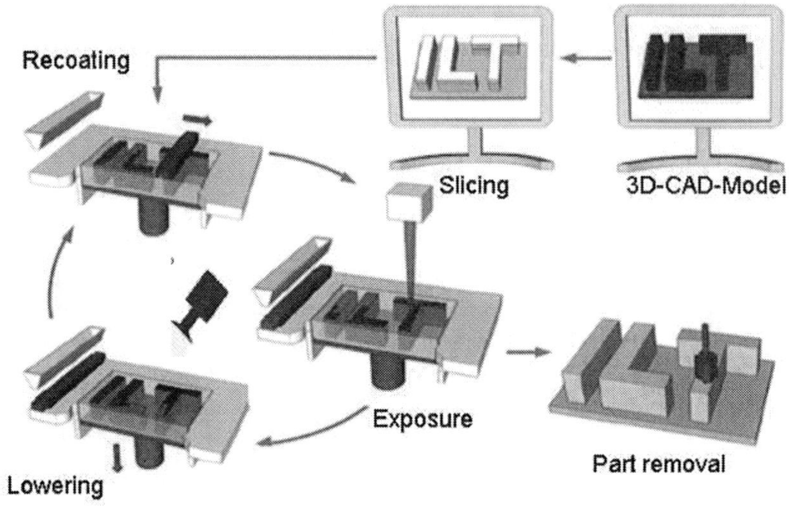

Fig. 1: Printing scheme with the location of the measurements in the printing cycle indicated. [3]

2.2 Numerical Simulation

The same parts as in the experiment and part of the platform were also used to calculate the temperature fields by simulation of the printing procedure using the finite element method. The grid density and intensity of the temperature source were adjusted so there would be a match with the experiment. A modified Oofem and Code-Aster solver code was used for the calculation.

Conclusion

Calibration and temperature measurements were made using a thermal imager during printing at the moment before the new layer was applied. These values were used to validate the results of the finite element simulation.

References

[1] T. DebRoy, H.L. Wei, J.S. Zuback, T. Mukherjee, J.W. Elmer, J.O. Milewski, A.M. Beese, A. Wilson-Heid, A. De, W. Zhang, Additive manufacturing of metallic components – process, structure and properties, Progress in Material Science 92 (2018) 112–224, https://doi.org/10.1016/j.pmatsci.2017.10.001.

[2] M. Montazeri, P. Rao, Sensor-based build condition monitoring in laser powder bed fusion additive manufacturing process using a spectral graph theoretic approach, Journal of Manufacturing Science and Engineering 140 (9) (2018) MANU-18-1023, https://doi.org/10.1115/1.4040264.

[3] US Research Nanomaterial, "Strain less steel powder" [online], available at: http://www.us-nano.com/inc/sdetail/14838. [Zugriff am 15 September 2017].

Comparison of Prediction Tools for Crack Growth under Variable Amplitude Loading

J. Šedek [1,a]

[1] VZLÚ – Czech Aerospace Research Centre, Beranových 130, 199 05 Prague - Letňany, Czech Republic
[a] sedek@vzlu.cz

Abstract: A crack growth prediction from two models based on the strip yield concept is presented. The FASTRAN model implemented in the AFGROW and the NASFLA model included in the NASGRO package were utilized. Models were used for crack growth under variable amplitude loading with the sequence characteristic for commuter airplanes. A retardation in crack growth is characteristic for such types of spectra. This effect was shown by the test conducted on middle-cracked tension specimens made from 7475 aluminum alloy. The results of prediction are not the same for both models. The differences of up to 50% compared to the test were demonstrated. The FASTRAN model was found to be sensitive to its parameters unlike the NASFLA.

Keywords: AFGROW; crack growth; FASTRAN; NASGRO; prediction.

1 Introduction

The crack growth prediction became more common due to the design of damage-tolerant aircraft. The interaction crack growth models are the most sophisticated, especially the strip yield model. Its concept brings a semi-analytical solution of the opening stress. Despite the fact that the concept was introduced in 80th of the last century by Newman [1], the models are used until today. Several implementations have been developed in laboratories worldwide, but FASTRAN in AFGROW [2] and NASFLA in NASGRO [3] become commercially available. Over the fact that both share the same basis, different results can be expected especially in situations when a large crack tip plastic zone develops.

2 Material Data

The material data for 7475 aluminum alloy were standard; $R_{p0.2} = 417$ MPa, $R_m = 486$ MPa and crack growth rate curve with Paris' parameters $C = 4.04e\text{-}10$ and $m = 2.85$. The specific ones are the Strip Yield Model data covering flat-to-slant regime of crack growth that is characterized by crack growth rates at the start and the end of this zone; da/dN_1 and da/dN_2 and corresponding constraint parameters α_1 and α_2.

Tab. 1: Strip Yield Model data.

Beginning of the flat-to- slant zone		End of the flat-to-slant zone	
α_1	2.3	α_2	1.2
Da/DN_1	2.00E-07	Da/DN_2	1.50E-06 8.00E-6

3 Experimental Finding

The 2 mm thick middle-cracked tension specimen M(T) made from 7475 aluminum alloy was used in an experimental evaluation. The crack was symmetrically initiated up to 2.5 mm from the saw-cut at the center hole. The commuter plane sequence was then used repeatedly up to the failure (see Rainflow of the sequence in Fig. 1). The retardation was observed from the crack length of 10 mm with approx. constant crack growth rate up to 30 mm before accelerated growth until the failure.

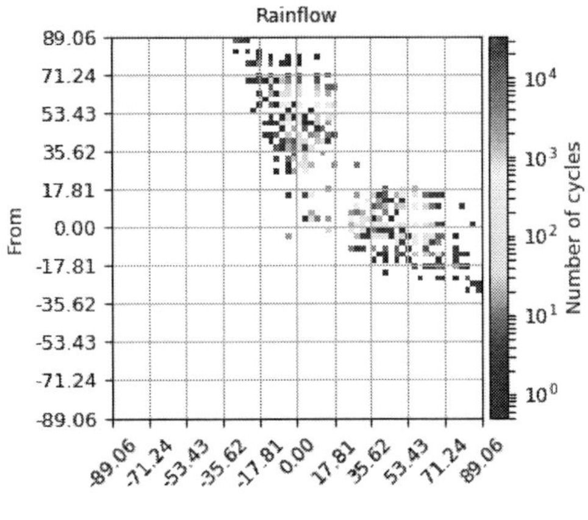

Fig. 1: Rainflow of the loading sequence.

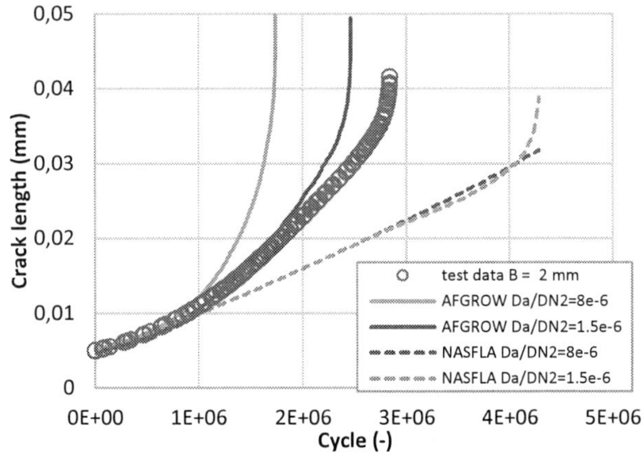

Fig. 2: Crack growth curves.

4 Prediction of Crack Growth

The prediction was made by Strip Yield Models implemented in AFGROW and NASGRO with the same input data. The prediction of crack growth is generally sensitive to a flat-to-slant regime. To demonstrate the sensitivity of both models, two values of Da/DN_2 were used.

The resulting crack growth curves are shown in Fig. 2 in comparison with test data. FASTRAN has shown to be sensitive to this parameter with Da/DN_2 = 1.5e-6 being closest to the experimental evaluation.

5 Conclusion

It is clear from the results, that the prediction is not the same for both models; FASTRAN and NASFLA The difference is up to 50% compared to the test. The sensitivity to parameters of a Strip Yield Model is demonstrated by varying Da/DN_2. The FASTRAN is sensitive unlike NASFLA. Using the same material data and SYM parameters, the AFGROW implementation is conservative in contrast to NASGRO. Therefore, both solutions should be treated with care with regard to application.

Acknowledgement

These works were carried out under the fund of the Technological Agency of ČR "National Competence Centre for Aeronautics and Space", partial project TN02000009/004 SeLLA.

References

[1] J. C. Newman Jr, A crack-closure model for predicting fatigue crack growth under aircraft spectrum loading, ASTM STP 748 (1981) 53–84.

[2] Fatigue & Fracture Associates, FASTRAN, A Fatigue Crack Growth Life-Prediction Code Based on the Crack-Closure Concept, Version 5.4, 2013.

[3] NASGRO V 10.2 [software], SRI, San Antonio, Texas, 2014.

Structural Health Monitoring of Composite Repairs using Ultrasonic Guided Waves

L. Šedková [1,a], O. Vích [1], I. Mlch [1], E. Horká [1]

[1] *VZLÚ – Czech Aerospace Research Centre, Beranových 130, 199 05 Prague - Letňany, Czech Republic*
[a] *sedkova@vzlu.cz*

Abstract: In this paper, adhesively bonded composite repair is subjected to multi-stage debonding/delamination. The ultrasonic guided wave (UGW) method is exploited for structural health monitoring (SHM) and ultrasonic A-scan for NDT validation of SHM results. It has been shown that A0 mode at 80 kHz is sufficiently sensitive to debonding/delamination using the time of flight (ToF) parameter. The specific extent of debonded/delaminated composite repair is possible to visualize using the probabilistic algorithm. The results obtained by the UGW method were in good agreement with precise ultrasonic detection.

Keywords: composite repairs; ultrasonic guided waves; structural health monitoring.

1 Introduction

Parts of damaged composite aircraft structures are usually replaced or repaired. In the case of composite repairs, manufacturing technology and the subsequent strategy for monitoring composite repairs need to be set. One option is to use adhesively bonded composite repairs.

The difficulty in predicting the long-term integrity of adhesive bonds limits the possibility of using the bonded patch repairs in aircraft structures. The adhesive bonding is susceptible to interfacial (adhesive) and bulk (cohesive) defects, making non-destructive inspection (NDI) essential to ascertain the quality of repairs.

Usually, conventional non-destructive inspection can only be applied at prescribed intervals when the aircraft is scheduled for servicing. Any increase in frequency required to detect disbonding may not be feasible or cost-effective. Structural health monitoring (SHM) offers an alternative approach in which the integrity of the patch system can be continuously, and possibly autonomously, monitored to enable the effective operational use of the structure with such a repair. SHM also has the advantage over most NDI techniques in that difficult-to-access places can be monitored readily once the system is installed.

2 Materials and Methods

Adhesively bonded composite repairs were tested in terms of debonding/delamination. The purpose of the test was to evaluate the SHM approach. The ultrasonic guided wave (UGW) method was selected for debonding/delamination monitoring using the Acellent Scan Genie II data acquisition system. Twelve PZT sensors placed around the composite repair were utilized for Lamb wave omnidirectional actuation. Sensors were mounted on the back side of the tested specimen. Actuation frequency was set to 80 kHz to actuate A0 mode and 130 kHz to actuate S0 mode. The reconstruction algorithm for probabilistic identification of damage (RAPID) was exploited for debonding imaging [1]. Precise debonding borderlines were marked by ultrasonic A-scans to verify results obtained by the UGW method.

3 Experimental Results

The evaluation is based on comparative measurements performed in different states. Measurements were performed for the baseline pristine state and eight delamination/debonding extents. The time of flight (ToF) of the corresponding peak of the wave packet was selected as a damage-sensitive parameter. Fig. 1a shows the ToF delay for A0 and S0 modes for an actuator-sensor path through the delamination region and through the pristine region. It can be seen that the A0 mode is significantly more sensitive to the presence of damage than the S0 mode. ToF delay is primarily caused by the propagation velocity change through the delaminated region.

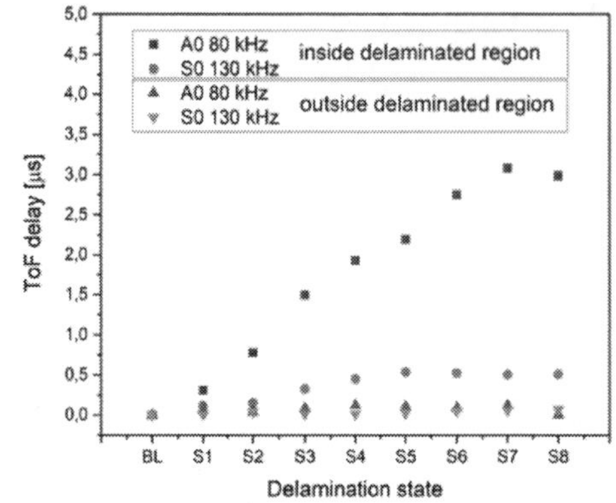

a)

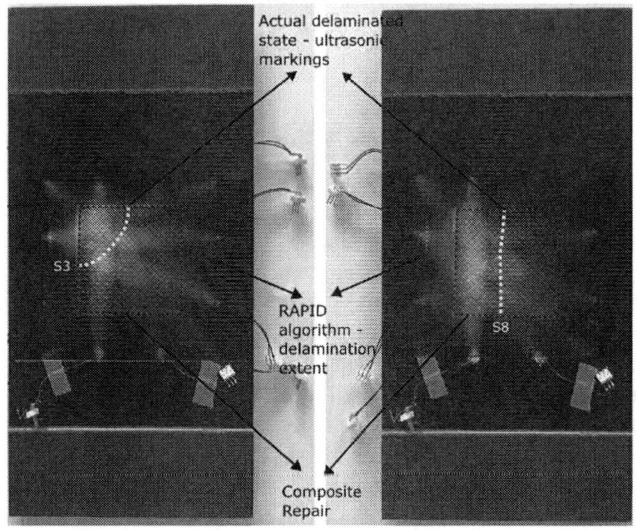

b)

Fig. 1: a) ToF delay of A0 and S0 mode inside and outside of the delaminated region, b) delamination extent – RAPID imaging compared to ultrasonic markings.

Based on the significant dependency of the A0 mode on the delamination extent and previous results [2], the ToF delay of A0 mode was exploited for evaluation using the RAPID algorithm. Fig. 1b shows delamination extent visualization in two states – S3 and the final S8. Both images correspond to the actual damage area determined by the ultrasonic A-scan.

4 Conclusion

Specimens with adhesively bonded composite repairs were subjected to delamination/debonding. The failure mechanism was determined to be delamination. Lamb waves were actuated using a network of 12 PZTs. ToF delay of A0 mode was exploited as a damage-sensitive parameter. Successful damage imaging was performed by means of probabilistic algorithm (RAPID).

Acknowledgement

The result was realized as part of the project solution number TN02000009 "National Competence Center for Aviation and Cosmonautics", sub-project TN02000009/05 "Smart, Modern manufacturing Processes".

References

[1] J. L. Rose, Guided Wave Imaging Methods, Ultrasonic Guided Waves in Solid Media, Chapter 21 – Guided Wave Imaging Methods, 402–420, 2014, https://doi.org/10.1017/CBO9781107273610.

[2] L. Šedková, J. Šedek, Defect detectability in composite plates with variable thickness using Lamb waves, in proc.: 58th International Scientific Conference Experimental Stress Analysis (EAN) (2020).

Methodologies for Modeling and Testing Bus Hybrid Structure Roll-Over Dynamics

S. Špirk [1,a], T. Kalina [1], J. Václavík [2]

[1] *University of West Bohemia, Czech Republic*
[2] *Research and Testing Institute Plzen, Czech Republic*
[a] *spirks@fst.zcu.cz*

Abstract: The realm of bus construction is a multifaceted and intricate domain, characterized by diverse requirements. Among these, the mass of the body frame stands out as a crucial parameter. A lighter body frame often translates to greater occupancy potential, yet it typically compromises the structural rigidity. However, striking a balance between reduced mass and maintained stiffness is achievable through geometric optimization or the integration of cutting-edge materials. This paper delves into a novel structural concept for bus construction, incorporating carbon fiber layers and foam fillings as localized reinforcements throughout the framework. The primary objective is to explore the viability of these hybrid joints within the bus construction landscape, encompassing hybrid beams and structural nodes.

Keywords: bus; hybrid; composite; foam; crash.

1 Introduction

The primary stress on the bus framework predominantly arises from fatigue and collision scenarios. While steel design offers lower raw material costs compared to composite alternatives, the manufacturing time for steel designs significantly exceeds that of composites.[1] This study hones in on crash dynamics, particularly focusing on roll-over scenarios (Ro.66), where composite materials and manufacturing techniques play a pivotal role in reducing vehicle weight for buses and trams.[2] Efforts to streamline bus construction, while maintaining or enhancing stiffness, unfold across two fronts. Firstly, establishing a numerical model, grounded in available material data, serves as the initial framework for experimental endeavors. Secondly, validating this numerical model entails cross-referencing it with empirical measurements. A prudent approach involves validating individual materials independently, followed by more complex assemblies, such as steel beams integrated with composite strips and foam. Though the core of the work revolves around a simplified beam model, the overarching findings will directly inform advancements in bus construction methodologies.

2 The Simplified Model of the Bus Segment

Given the unique nature of certain materials and their applications in bus construction, the task at hand demands a substantial amount of experimentation. Traditional material testing methods, such as tensile testing alone, fall short in providing comprehensive insights. Instead, the adoption of three-point and four-point bending tests emerges as a viable approach, mirroring methodologies documented in relevant literature. [3]

These tests hold promise due to their applicability to samples resembling actual bus constructions and their deformation patterns akin to those observed in rollover tests.

The process of developing material models unfolds gradually. Steel components of the structure, for instance, can be characterized based on standard tensile tests. However, the validation of the finite element method (FEM) material model necessitates more nuanced assessments. The beam test, representing the foundational element of the bus structure and significantly influencing rollover test behavior, stands as a pertinent validation point for the FEM material model (refer to Fig. 1). Alignment between deformation characteristics serves as a litmus test for the adequacy of the steel material model. Notably, the numerical reliability of the FEM model is gauged through an energy balance criterion, where the total energy change must remain below 10%.

For computations pertaining to passive safety, the project leverages the FEM solver Pam-Crash, integrated within the VPS (Virtual Performance Solution) software suite. This solver facilitates the analysis of structural dynamics employing an explicit integration scheme, ideally suited for scenarios characterized by significant nonlinearities, be they geometric, material, or contact related. Constructing the calculation framework entails the creation of a FEM mesh encompassing requisite entities such as welded joints, glued joints, and contacts. Crucially, the development of relevant material models assumes paramount importance.

Innovative design endeavors necessitate the utilization of diverse material models, which can be categorized into several key classifications: Standard steel materials, Composite materials, Foam, and Glue. Each category plays a distinct yet integral role in shaping the structural integrity and performance characteristics of the bus construction under scrutiny. The first results of numerical simulations can be seen in Fig. 1.

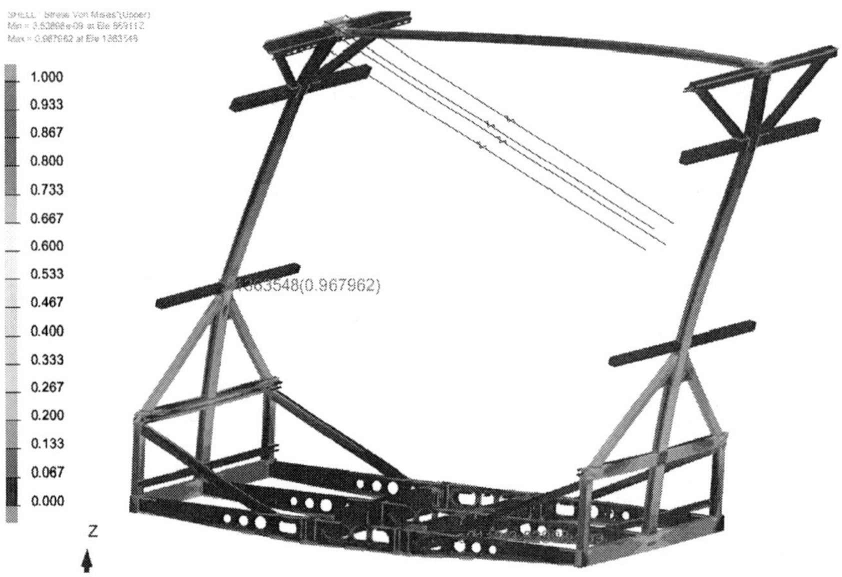

Fig. 1: The Von Mises stress, segment containing CFRP strips only on the sides.

3 Conclusion

The paper presents preliminary findings and methodologies for numerically simulating hybrid beam structures. Initial insights are gleaned from a partially validated model featuring a steel hollow profile, augmented with a composite strip and foam filling. Observations indicate that optimal performance is achieved when the composite material closely matches the Young´s modulus of steel, highlighting the importance of effective collaboration between composite and steel components. The integrity of the glue connection emerges as a critical factor in ensuring structural robustness. Validation of material models can be augmented through three-point bending tests and their corresponding simulations. However, for comprehensive validation, it is recommended to conduct numerical simulations alongside full-scale experiments on bus segments. While such experiments are slated for the future, they promise to offer invaluable insights into the performance and behavior of the hybrid beam structure under real-world conditions.

Acknowledgement

This work has been supported by the project TH71020003 "Composite reinforcement in a light stainless steel bus structure" provided by the Ministry of Industry and Trade of Czech Republic.

References

[1] E. Larodé et al., A new concept of a bus structure made of composite materials by using continuous transversal frames, Composite Structures 32 (1995) 345–356.

[2] K. Hee-Young et al., A study on the crashworthiness and rollover characteristics of low-floor bus made of sandwich composites, Journal of Mechanical Science and Technology 23 (2009) 2686–2693.

[3] M. Kriescher et al., Possibilities for the use of metal-hybrid-structures for vehicle crash load cases, in proc.: 1st International Conference Euro Hybrid Materials and Structures (2014), Stade, Germany, 10–11.

Measurement of Residual Stresses

J. Václavík [1,a], R. Musilová [1]

[1] *Research and Testing Institute Plzen, Tylova 1581/46, 301 00 Plzeň, Czech Republic*
[a] *vaclavik@vzuplzen.cz*

Abstract: This article contains an overview of methods for measuring macroscopic residual stresses using semi-destructive and non-destructive techniques, including some applications.

Keywords: residual stresses; hole-drilling method; ultrasonic method; magnetic method.

1 Introduction

Residual stress is an important parameter in respect of the failure of the structure or mechanical component.

Residual stresses are presented in the material without external forces and are one of the parameters of the overall condition of the material. They arise as a result of technological production procedures and have an effect on its strength, fatigue or corrosion characteristics.

In general, compressive residual stresses are advantageous (they increase the fatigue limit and extend the life of components) and tensile residual stresses are disadvantageous, they can significantly lower the fatigue strength of the structure and thus increase the structure service or residual service life, cause the formation of cracks, stress corrosion cracking, reduction of the fatigue limit or brittle fracture resistance.

At high values of residual stresses, undesirable deformations due to their release and degradation of mechanical properties can occur not only in the case of tensile but also compressive stresses.

Macroscopic residual stresses (type I) are homogeneous over areas of an order of magnitude larger than the grain size. These stresses are connected to the external load of the body and affect its strength properties. Their change or loosening is associated with macroscopic changes in the shape of the part.

Microscopic residual stresses (type II) represent the average deviation from the macroscopic stress at a given location. Submicroscopic residual stresses (type III) are defined as a local deviation from the average microscopic stress.

Several technics are available for subsurface residual stress measurement such hole-drilling, ring-core, slitting and X-Ray, ultrasonic or magnetic methods [1], see Fig. 1.

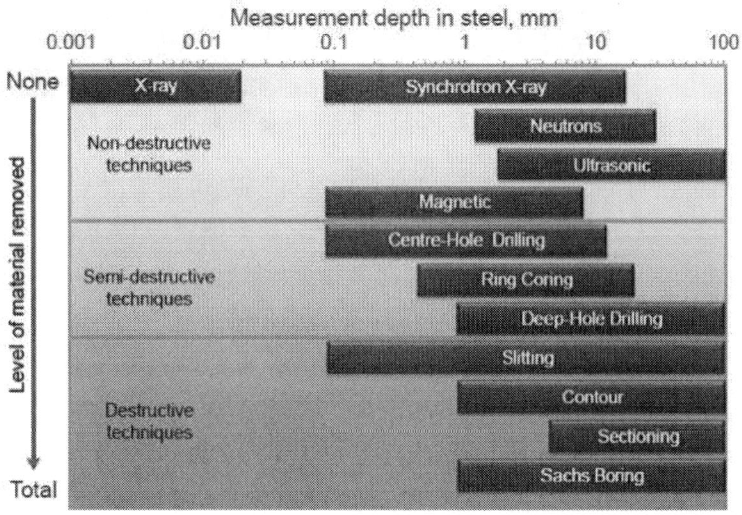

Fig. 1: Methods of residual stress measurement (weqter.co.uk).

Mechanical methods are based on the measurement of the stress change when the existing residual stresses are released or when residual stresses are introduced during the given technological process. Deflection methods, boring method, division methods and indentation methods are used. Diffraction methods are used to measure residual stress in crystalline materials, which use specific properties of radiation scattering on the crystal lattice. X-rays, synchrotron radiation, neutron radiation and electron beam are used. Methods using the physical properties of materials are based on the change of some specific properties of the material depending on the stress. Electromagnetic, ultrasonic, thermomechanical or photo elastic techniques are used.

2 Results

In this contribution, some investigations of residual stresses performed by Research and Testing Institute Plzen are described. All these examples cover the macroscopic residual stresses. It deals especially of semi-destructive hole-drilling, groove grinding and ring-core methods for residual stress measurement on railway axles, large shafts and forgings, magnetostriction method and ultrasonic method and some special usage of these methods for measurements on plasmatic sprayed layers.

a) Railway axle with the drilling jig

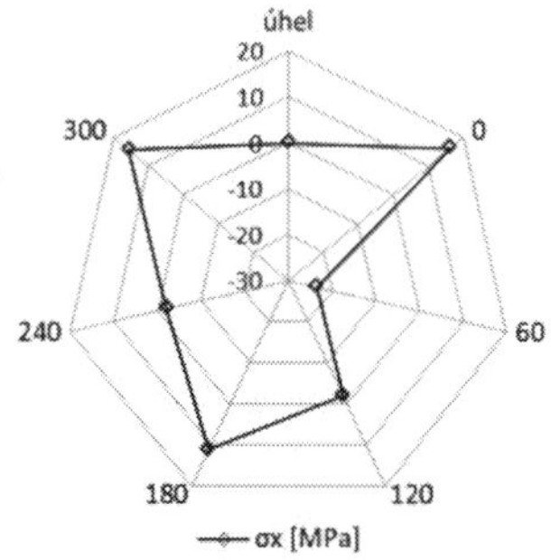

b) Angular distribution of axial stresses in 2mm depth of the axle

Fig. 2: Example of residual stress measurement in the middle of railway axle, using hole drilling method.

Fig. 3: Measurement of longitudinal residual stresses using subsurface ultrasonic waves.

Fig. 4: Investigation of residual stresses on turbine shaft using magnetostriction method.

3 Conclusion

For various tasks, it is always necessary to determine not only a suitable method for measuring residual stresses, but also to determine the permissible value depending on the material, the shape of the part and the operating stress.

Acknowledgement

The article originated in the framework of the institutional support for the long-term conceptual development of the research organization.

References

[1] G.S. Schajer, Practical residual stress measurement methods, first ed., John Wiley & Sons Ltd, United Kingdom, 2013.

Biaxial Cruciform Testing of Fiber Composite Materials

V. Vomáčko [1,2,a], M. Šulc [2], J. Kolář [2]

[1] *Faculty of Mechatronics, Informatics and Interdisciplinary Studies, Technical University of Liberec, Studentská 1402/2, 46117, Liberec, Czech Republic*

[2] *Department of Virtual and Applied Mechanics, VÚTS, a.s., Svárovská 619, 46001, Liberec, Czech Republic*

[a] *vaclav.vomacko@tul.cz*

Abstract: The biaxial cruciform testing method was adopted to validate the failure criteria results obtained through numerical simulations. A biaxial testing machine was developed, and carbon fiber-epoxy laminate specimens were tested. Failure strains were evaluated in the central area of the specimens. Specimens without tabs failed in the arm section, while specimens with glass-fiber tabs failed in the central section. However, both failed prematurely, as confirmed by numerical simulations with a progressive damage model, indicating further need for the optimization of the cruciform specimen.

Keywords: biaxial; cruciform; composite; failure; validation.

1 Introduction

Different approaches can be employed to validate the safety limits of parts made from fiber composites, which are obtained from numerical simulations. Traditionally, this involves prototyping and conducting experiments to simulate real-world conditions. Alternatively, biaxial cruciform tests can be employed, following these steps: i) evaluation of the critical stress state in the simulation, ii) manufacturing of a cruciform specimen using the same material and layup as is in a critical area determined by simulation, iii) induction of the critical stress state in the specimen by biaxial test machine and iv) comparison of the failure stresses and strains obtained from both, the test and simulations.

The advantage of performing biaxial tests is that specimens are made from flat laminates, resulting in a relatively fast preparation compared to prototype experiments. The same instrumentation, measurement technique and evaluation are used for all biaxial tests.

In the case of prototype testing, the prototype part has to be manufactured which comes at the expense of high costs. On the other hand, prototype testing is a robust approach for the validation of simulations, resulting in trustworthy results.

2 Biaxial Cruciform Tests

Biaxial cruciform test machine was developed at VÚTS, a.s. (see Fig. 1). The test machine consists of 4 independent actuators with a maximal load capacity of 10 kN. The stroke of the machine is 350 mm, which allows composite and elastomer testing. Tests can be performed in both displacement and load control modes. Displacements and strains are measured by the Digital Image Correlation system Monet 3D.

Equibiaxial tensile tests were performed in the displacement control mode. The cruciform specimen geometry features a double corner fillet with reduced thickness in the central area. Laminates were produced by vacuum infusion process using LH385 epoxy system. Specimen were milled to final shapes by CNC machine and GF tabs were bonded to the CF laminate using Letoxit PL20, see Fig. 1. Carbon-fiber laminate of interest featuring cross-ply layup $[0/90]_S$ and three different tab types are examined: i) no tabs, ii) glass-fiber laminate tabs $[\pm 45F]_3$ and iii) glass-fiber laminate tabs $[0F]_3$.

A quarter numerical model was employed to simulate biaxial testing of cruciform specimens (see Fig. 1). The simulations were performed using the finite element software Ansys 2022R2. The boundary conditions were $u = 0.5$ mm in the end of the tabs. The material model utilized in this study is orthotropic elasticity with a progressive damage approach.

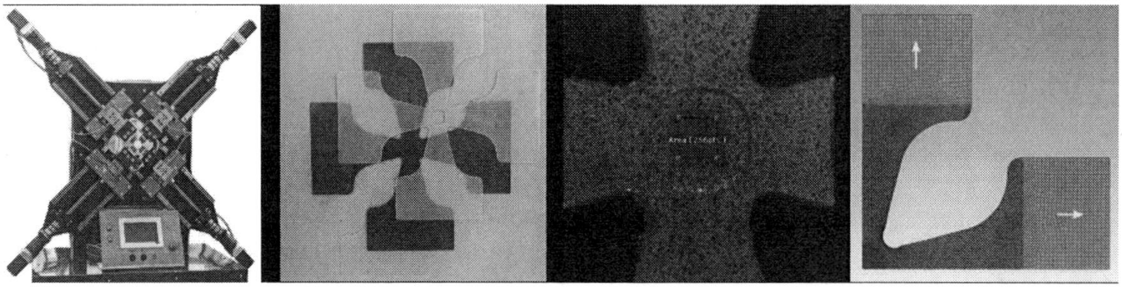

Fig. 1: Testing machine, specimen with tabs, strain evaluation area at DIC and mesh with BC.

3 Results and Discussion

Strains were evaluated in central square 8x8 mm of the specimen, see Fig. 1. Average values of measured failure strains are summarized in Tab. 1. The specimens set without tabs demonstrate consistent results with minimal measured value ϵ_f= 0.95 % and maximal measured value ϵ_f= 1.15 %. Both specimen sets with GF tabs show similar behavior reaching higher failure strains from (ϵ_f= 1.27 % to ϵ_f= 1.6 %) compared to the specimens with no tabs. The value of the highest reached failure strain ϵ_f= 1.6 % is close to the value $\epsilon_{f\ uniaxial}$ = 1.65 % obtained by uniaxial tests on coupon specimens of the same layup and material. The results show that the use of tabs significantly enhances the values of failure strain.

The failure observed in specimens, as depicted in Fig. 2, indicates successful biaxial tests for specimens with GF tabs. In contrast to specimens with no tabs, where failure occurs in a single arm.

However, the area of failure near the transition between tabs and the central section corresponds with the maxima of the Puck failure criteria observed in simulations. According to the simulations, the failure initiates in the 0 fiber direction, see Fig. 2. It can be explained by stiffness change and therefore stress concentration in the transition from tabs to the central area, where the outer plies carrying load are more loaded than inner plies in second direction - which are protected by outer plies and load transition is smoother. This observation implies that the failures are still occurring prematurely and therefore there is a further need for the optimization of the cruciform specimen.

Tab. 1: Measured average values of failure strains.

Experiment	ϵ_{fx} [%]	ϵ_{fy} [%]
No tabs	1.08	1.06
GF $[45F]_3$	1.35	1.44
GF $[0F]_3$	1.37	1.51

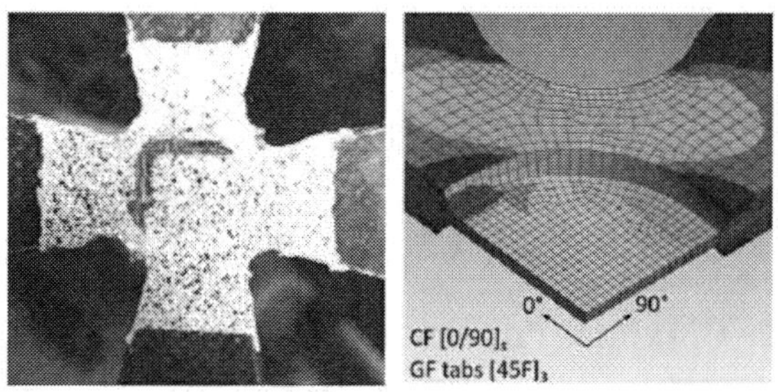

Fig. 2: Failure of specimen and maximum failure index near ultimate failure.

Acknowledgement

This publication was supported by the Ministry of Industry and Trade (MPO) within the framework of institutional support for long-term strategic development of the research organization - provider MPO, recipient VÚTS, a.s. This work was also supported by the Student Grant Scheme at the Technical University of Liberec through project Nr. SGS-2023-3378.

Optimization of Residual Stresses in 3D Print – HVOF Sprayed Coatings

M. Vostřák [1,a], I. Zetková [2], R. Musilová [1], J. Václavík [1], M. Křížek [2], Š. Houdková [1]

[1] *Research and Testing Institute Plzen, Tylova 1581/46, 30100 Plzeň, Czech Republic*
[2] *University of West Bohemia, Univerzitní 8/2732, 301 00 Plzeň, Czech Republic*
[a] *vostrak@vzuplzen.cz*

Abstract: The integration of 3D printing in manufacturing offers unique opportunities for producing customized parts but presents challenges in ensuring optimal surface properties. This study is part of a project focused on investigating the enhancement of surface properties and fatigue through HVOF thermal spraying of hard metal coatings on 3D printed metal substrates. The goal of this work is to analyze residual stresses – in the applied coating and the 3D printed substrate. By adjusting production parameters, significant reductions in tensile residual stresses were achieved, indicating potential advancements in material fatigue resistance for additive manufacturing applications.

Keywords: HVOF; DMLS 3D printing; hard metal coatings; residual stress; fatigue.

1 Introduction

3D printing provides the possibility of producing spare unique parts or very fragmented parts with an optimized shape. However, due to the nature of the production of parts by additive technology, some material properties do not correspond to the properties of parts obtained by conventional technologies. It concerns especially the fatigue properties of 3D printed parts.

One of the investigated options to increase fatigue life is to provide the surface of the material with thermal spraying. In the frame of solved project, the effort is the development of a 3D printed metal substrate system – HVOF sprayed coatings with increased resistance to aggressive operating environments with a predictable response to possible dynamic, cyclic and vibration loads.

Attention is paid in particular to the compatibility of the laser 3D printing and thermal spraying process and monitoring the resistance of the developed 3D printing-thermal spraying system to dynamic impact loads and the liveliness of the system during cyclic loading. An integral part is also the development of evaluation methodologies for dynamic impact loading and cyclic loading with regard to the composite nature of 3D printing-incandescence spraying systems.

One of the parameters for monitoring the effectiveness of fatigue resistance is the level of residual stresses, which is the important parameter in respect of the failure of the structure or mechanical component.

2 Solution

As part of the project solution, parameters for injection molding and development of 3D printing procedures were developed. The printing parameters of tool steel W-Nr.1.2709 with MS1 powder trademark for elimination of residual stress were investigated. It was possible to reduce the values of residual stresses into the compressive area. The reduction of residual stresses was achieved by increasing the preheating from the standard temperature t = 40 °C (designation SP) to 120 °C (designation SP 120) during the melting process.

The used composite mixtures of carbide particles with a metal matrix have considerable wear resistance and a tough matrix and are referred to as "hard metal" or "metal-ceramic". By far, these compounds are based on tungsten carbide (WC) and chromium carbide (Cr3C2) in combination with matrices containing nickel, cobalt and chromium. Investigated materials were Amperit 543.074 (42WC-42Cr3C2-16Ni) and Woka 7502 (37WC-43Cr3C2-18NiCo).

Residual stresses were analyzed on 3D printed blocks with two levels of preheating without and with applied two kinds of coatings.

Residual stress measurement was performed by using hole-drilling method using Vishay RS-200 device with air turbine with the speed above 200 000 rpm. As a strain gauge rosette TML FRS-2-11 was used and as a tool the diamond cutter SINT D2/1 (ϕ1.6 mm). For the data acquisition the 8-chanell amplifier HBM SPIDER8 and software Catman Professional were used. The evaluation of residual stresses was performed by using the software SINT EVAL7.2 with integral method. The measurement was performed independently by two laboratories, (see Fig. 1).

The profiles of the residual stresses along the depth from the surface are documented in Fig. 2. The reduction of tensile residual stresses after preheating the sample is documented in Fig. 2a. The comparison of residual stresses profile on 3D printed samples with Amperit or Woka layers is performed in Fig. 2b.

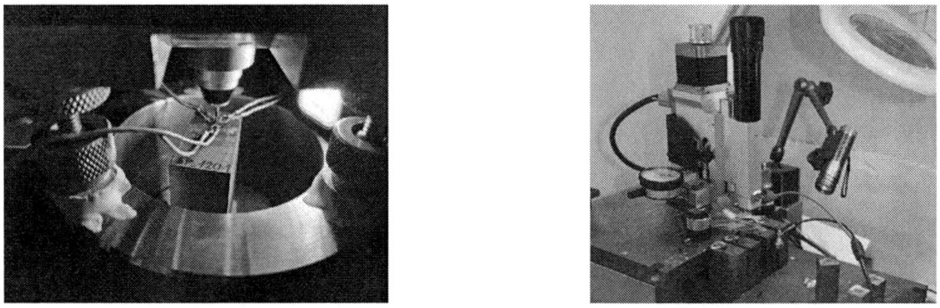

a) detail of the attachment of the measured sample and the measuring fixture Vishay RS-200 (RTI)

b) measuring device HBM Restan for automatic measurement of residual stresses (UWB)

Fig. 1: Devices for residual stress measurement from two laboratories.

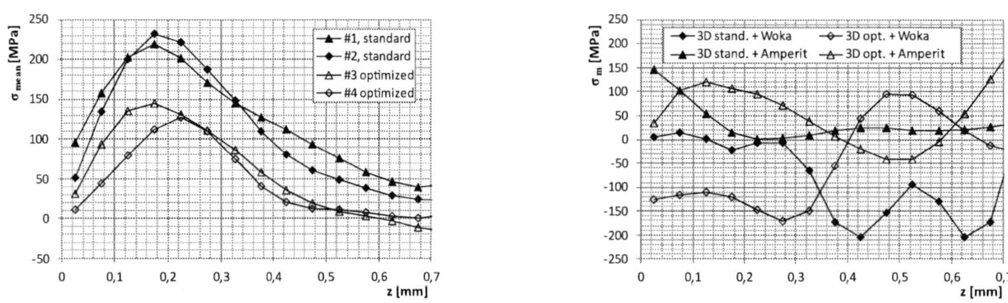

a) samples as printed without layer

b) printed samples with Woka or Amperit layer

Fig. 2: Mean component of residual stresses for standard (t = 40 °C) and optimized (t = 120 °C) preheating.

3 Conclusion

For steel produced by 3D printing, with the optimization of production parameters, especially by increasing the preheating temperature, substantially reduce tensile residual stresses. The sample, produced by 3D printing with Woka 7502 layer, has compressive stresses on the surface, which in the base material turn into tensile values. The sample, produced by 3D printing with Amperit 543.074 layer, has tensile stresses under the surface, but does not change significantly at the transition of the layer to the basic material.

Acknowledgement

This work was supported by M-ERA.Net 2 Call 2020 with the name "Development of 3D print-thermal spray" systems for applications with dynamic and impact loading Epsilon under the name TH75020003.

References

[1] ASTM Determining residual stresses by the hole-drilling strain-gage method. Standard test method E837-20. American Society for Testing and Materials, West Conshohocken, PA (2013).

[2] R. Musilová, M. Vostřák, K. Trojan, Residual stress measurement measured by different type of method on the 3D print-thermal spray, in proc.: 61st conference on experimental stress analysis EAN (2023), book of extended abstracts, eds. M. Hagara, E. Huňady, Technical University of Košice, Košice, Slovakia, 56–57.

Efficiency of Reinforcement Materials on Bus Frame Stability

V. Yanovych [1,2,a], J. Václavík [1]

[1] *Research and Testing Institute Plzen, Dynamic Testing Laboratory, Czech Republic*
[2] *Department of Power System Engineering, Faculty of Mechanical Engineering, University of West Bohemia, Czech Republic*
[a] *yanovych@vzuplzen.cz*

Abstract: This paper presents the results of an experimental investigation of the CFRP and epoxy foam's effect on bus frame stability. As part of the research, a series of quasi-static tests were conducted to analyze the behavior of the test specimen in four experimental scenarios. Initially, the steel structure was tested, followed by reinforcement with carbon fiber on one post, then extending the carbon fiber reinforcement to both the crossbar and side posts. The final scenario involved filling the steel structure with epoxy foam, excluding carbon fiber. Generally, according to the results obtained, the reinforcing materials have a significant impact on the structural stability of the specimen structure. Namely, the deformation energy in the case of epoxy foam and CFRP reinforcement decreased by about 15 %.

Keywords: bus frame; reinforcement materials; structural stability; deformation.

1 Introduction

Bus accidents occur frequently worldwide. In Europe, approximately 150K individuals sustain injuries, while 150 lose their lives annually due to such incidents [1, 2]. Consequently, prioritizing the safety of bus passengers has become a critical issue in the bus industry, reflecting the escalating safety challenges seen in the automotive sector. Safeguarding the physical safety of bus passengers isn't solely a matter of marketing; it's also a duty aligned with international standards.

Currently, one of the most promising methods of increasing the structural stability of passenger vehicles is the application of carbon fiber-reinforced polymer (CFRP) and epoxy foam [3–5]. A large number of different studies have been devoted to this issue. However, the reinforcement process is usually adapted to the specific characteristics of each structure, resulting in a customized investigation of each experimental case.

2 Test Configuration and Measuring Equipment

Experimental investigations were conducted at the Dynamic Testing Laboratory of the Research and Testing Institute Pilsen (Czech Republic). The primary focus of the study was to assess the effectiveness of various reinforcing materials on the structural stability of the frame in passenger transport, specifically in the event of a bus rollover accident. In such situations, the roof and sidewalls undergo significant deformation, which can lead to a critical reduction in safe space, resulting in injury to passengers.

In the current study, a standard bus frame section was used as the test specimen. The experimental plan included conducting a series of quasi-static tests to analyze the behavior of the test specimen across four experimental scenarios. Initially, the investigation focused on a structure comprising solely of steel. After that, in the second scenario, the frame received reinforcement by carbon fiber exclusively on one post. Moving on to the third case, carbon fiber reinforcement (CFRP) was extended to cover both posts and the crossbar of the frame. It is crucial to note that in this configuration, the upper reinforcing lamella was positioned between welds. Finally, the fourth scenario involved filling the steel structure with epoxy foam, without adding carbon fiber. Figure 1 illustrates the organization of the test bench for assessing the structural stability of the specimen under various scenarios. In all experimental scenarios, a force of 10 kN was applied at 22 degrees relative to the horizontal plane. To quantify the deformation pattern of the specimen, a variety of measurement sensors were employed.

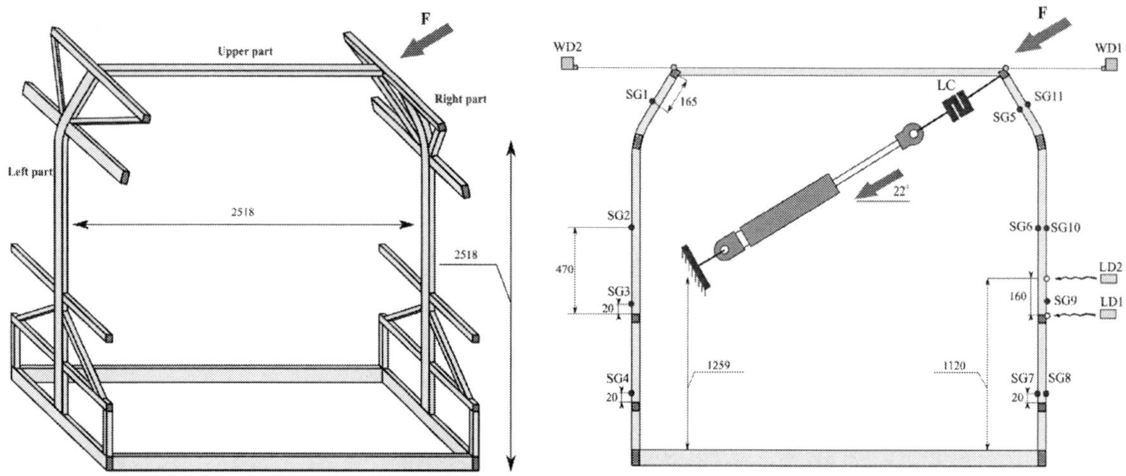

Fig. 1: Experimental setup, there a test specimen sketch on the left and a test bench on the right. Where SG - strain gauges, WD - wire digital sensors WDS-1000-P60, LD - laser distance sensors HG-C1200, LC - load cell HBM S9M/20kN.

3 Experimental Results

The experimental results show that in all test scenarios, the external load causes clear localized deformation zones that resemble bulges at the joints of the upper and side frame elements. It is notable that the orientation of the bulges differs depending on the side of the specimen. Specifically, the bulge on the right side displays a downward curvature, whereas on the left side, it curves upward. This contrast is attributed to the compression applied by the hydraulic cylinder on the right side. As a result, the greatest stress and subsequent plastic deformation of the structure are observed in this area.

Figure 2 depicts the force-displacement relationship for all experimental scenarios. Notably, the maximum deformation force is evident for specimen reinforced with carbon fiber on both the crossbar and side posts. Additionally, the obtained results exhibit a significant data perturbation at 22 seconds, which is attributed to the weld failure. It's worth noting that for all experimental cases, except for epoxy foam, the deformation curves exhibit clearly defined yield points. That indicates the elastic behavior of the specimen construction in this case. In contrast, in the foam scenario, a more pronounced plastic behavior is observed.

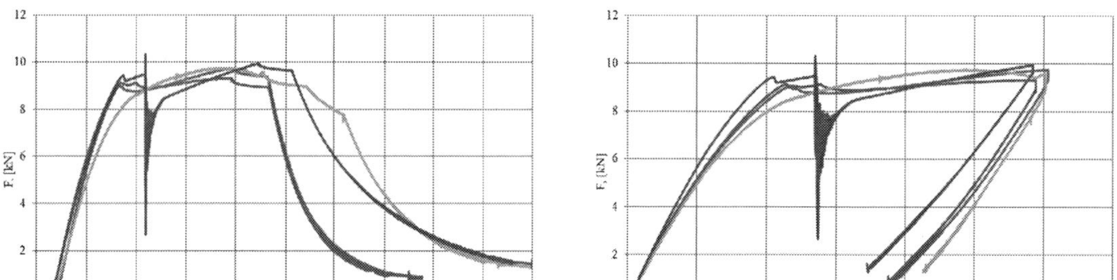

Fig. 2: Force-deformation dependence for all experimental scenarios.

Additionally, the elastic deformation energy of the specimen was estimated based on the obtained data, considering the type of reinforcement. Calculation data in Fig. 3 showed that in the cases of epoxy foam and carbon fiber reinforcement on the crossbar and side posts, the deformation energy decreased by approximately 15%. In general, reinforcing materials have a significant impact on the structural stability of the sample structure. This effect is particularly evident in Fig. 4, which represents the strain estimation at measurement point 9. It can be observed that the foam and carbon fiber reinforcement led to approximately a 65% and 48% reduction in strain, respectively.

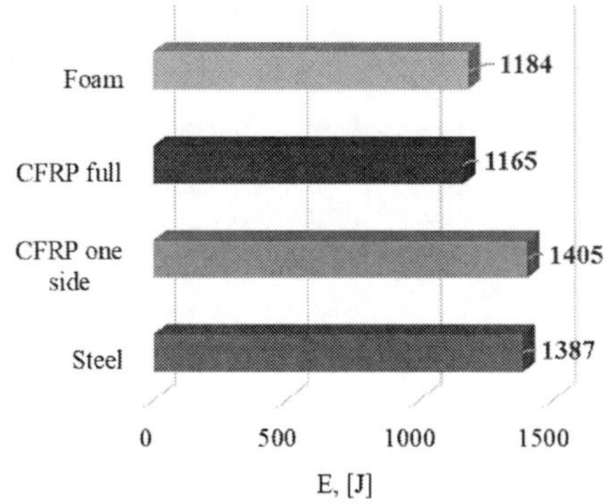

Fig. 3: Elastic deformation energy depending on the bus frame reinforcement type.

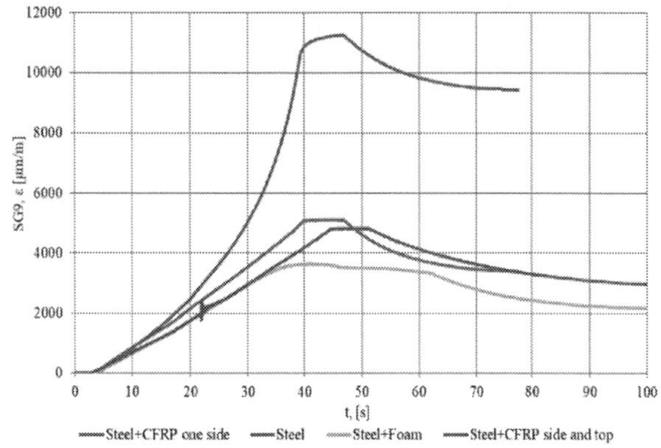

Fig. 4: Strain-time distribution at measurement point 9 depending on type reinforcement.

4 Conclusion

In the classical case of a frame without reinforcement, the structure undergoes significant plastic deformations due to the applied load. In contrast, the epoxy foam acts as an absorber of the deformation energy and helps to effectively distribute the resulting stresses evenly throughout the frame structure. Therefore, the plastic deformation on both sides of the frame is similar. This statement is also confirmed by the configuration of the resulting bulges, which are much wider than in other reinforcement test cases. At the same time, carbon fiber reinforcement increases the stiffness of the frame by effectively redistributing the stress state of the structure. However, it is important to emphasize that, unlike epoxy foam, carbon fiber does not absorb energy, but only redistributes it. This distinct feature causes a localized concentration of significant stress at the welded joints of the frame. Consequently, this concentration leads to the failure of the welded joint outside the carbon fiber reinforcement. Generally, it can be concluded that carbon fiber and foam reinforcement are effective approach to improving the stability of bus frame structures. Namely, their application reduced the deformation energy by almost 15%.

Acknowledgement

The research was performed within the framework of the M-ERA.NET 2019 competition and was supported by the Technology Agency of the Czech Republic under the number TH71020003.

References

[1] P. Albertsson, T. Falkmer, Is there a pattern in European bus and coach incidents? A literature analysis with special focus on injury causation and injury mechanisms, Accident Analysis and Prevention 37(2) (2005) 225–233, https://doi.org/10.1016/j.aap.2004.03.006.

[2] K. Kyoung-Tak et al., Design of a composite roll bar for the improvement of bus rollover crashworthiness, Composites Part B: Engineering 43 (2012) 1705–1713, http://dx.doi.org/10.1016/j.compositesb.2012.01.072.

[3] A. Pawlak et al., The Use of CFRP for Structural Reinforcement-Literature Review, Metals 12 (2022) 1470, http://dx.doi.org/10.3390/met12091470.

[4] R. Sen, L. Liby, A. Mullins, Strengthening steel bridge sections using CFRP laminates, Composites Part B: Engineering 32 (2001) 309–322, http://dx.doi.org/10.1016/S1359-8368(01)00006-3.

[5] J.G. Teng et al., Behaviour and strength of FRP-strengthened RC structures: A state-of-the-art review, Proceedings of the Institution of Civil Engineers-Structures and Buildings 156 (2003) 334–335, http://dx.doi.org/10.1680/stbu.156.3.334.37862.

Index of Authors